Perturbations in Mechatronic, Robotic, and Energy Systems

Perturbations in Mechatronic, Robotic, and Energy Systems

Estimation and Attenuation

Jose de Jesus Rubio

Instituto Politécnico Nacional
ESIME Azcapotzalco
Sección de Estudios de Posgrado e Investigación

WILEY

Published by John Wiley & Sons, Inc., Hoboken, New Jersey.

Library of Congress Cataloging-in-Publication Data

Names: Rubio, Jose de Jesus author

Title: Perturbations in mechatronic, robotic, and energy systems : estimation and attenuation / Jose de Jesus Rubio.

Description: Hoboken, New Jersey : Wiley, [2026] | Includes bibliographical references.

Identifiers: LCCN 2025054157 | ISBN 9781394380633 hardback | ISBN 9781394380640 epub | ISBN 9781394380657 adobe pdf | ISBN 9781394380664 ebook

Subjects: LCSH: Perturbation (Mathematics) | Engineering mathematics | Attenuation (Physics) | Control theory

Classification: LCC QA871 .R895 2026
LC record available at https://lccn.loc.gov/2025054157

Cover Design: Wiley
Cover Image: © FullFrames/stock.adobe.com

SKY10147931_022526

Contents

Preface *xi*
Acknowledgments *xiii*
Introduction *xv*

1 Geometric Method and Structure at Infinity Regulators for the Perturbation Attenuation *1*
1.1 Introduction *1*
1.2 Notation *2*
1.3 Perturbation Attenuation Problem *2*
1.4 Perturbation Attenuation Via Geometric Method Regulator *3*
1.5 Perturbation Attenuation Via Structure at Infinity Regulator *6*
1.6 Mechanical Process *8*
1.6.1 The States-space Process *8*
1.6.2 Perturbation Attenuation Via the Structure at Infinity Regulator *10*
1.6.3 Perturbation Attenuation Via the Geometric Method Regulator *11*
1.6.4 Simulation *13*
1.7 Thermal Process *15*
1.7.1 Perturbation Attenuation Via the Structure at Infinity Regulator *17*
1.7.2 Perturbation Attenuation Via the Geometric Method Regulator *18*
1.7.3 Simulation *19*
1.8 Concluding Remarks *22*
References *22*

2 Structure Regulator for the Perturbation Attenuation in Two Processes *25*
2.1 Introduction *25*
2.2 Structure at Infinity Strategy for the Perturbation Attenuation *26*
2.3 Structure Regulator for the Perturbation Attenuation *27*
2.3.1 Regulator Design *27*
2.3.2 Stability Analysis *29*
2.3.3 Convergence Analysis *29*
2.4 Perturbation Attenuation in a Three-phase Electric Circuit *30*
2.4.1 Structure at Infinity Regulator Design *33*
2.4.2 Structure Regulator Design *33*
2.4.3 Comparisons *35*
2.5 Perturbation Attenuation in a Plotter *38*
2.5.1 Structure at Infinity Regulator Design *40*
2.5.2 Structure Regulator Design *40*
2.5.3 Comparisons *41*
2.6 Concluding Remarks *45*
References *45*

3 Hybrid Regulator with Observer for the Estimation and Attenuation of Perturbations *49*
3.1 Introduction *49*
3.2 The Perturbed Process *51*
3.3 Hybrid Observer *51*
3.3.1 Observer Design *51*
3.3.2 Stability Analysis *52*
3.3.3 Estimation of the Perturbations *53*
3.4 Hybrid Regulator with Observer *53*
3.4.1 Regulator Design *53*
3.4.2 Stability Analysis *54*
3.5 Plotter Process *56*
3.6 Suspension Process *64*
3.7 Concluding Remarks *72*
References *72*

4 Sliding-mode Regulator for the Perturbations Attenuation in Two Tank Processes *75*
4.1 Introduction *75*
4.2 The Perturbed Process *76*
4.3 The Sliding Mode Regulator *77*
4.3.1 The Regulator Development *77*

4.3.2 The Stability Analysis *77*
4.4 The Thermal Process *79*
4.5 The Hydraulic Process *84*
4.6 Concluding Remarks *89*
References *90*

5 Uniform Stable Observer for the Perturbation Estimation in Two Energy Processes *93*
5.1 Introduction *93*
5.2 Observer for the Processes Without Perturbations *94*
5.2.1 Observer Design *95*
5.2.2 Stability Analysis of the Observer *95*
5.3 Observer for the Processes with Perturbations *96*
5.3.1 Observer Design *96*
5.3.2 Stability Analysis of the Observer *97*
5.3.3 Convergence Analysis of the Observer *98*
5.3.4 Estimation of the Perturbations *98*
5.4 Wind Turbine *99*
5.4.1 Comparisons *103*
5.5 Electric Vehicle *108*
5.5.1 Comparisons *111*
5.6 Concluding Remarks *114*
References *115*

6 The Perturbation Estimation in Two Gas Processes *117*
6.1 Introduction *117*
6.2 The Estimator for the Variable and Perturbation Estimation *119*
6.2.1 The Variable Estimator *119*
6.2.2 The Convergence Analysis of the Variable Estimator *120*
6.2.3 The Perturbations Estimator *120*
6.3 The Gas Turbine Process *123*
6.3.1 Comparisons *125*
6.4 The Gasification Process *129*
6.4.1 Comparisons *132*
6.5 Concluding Remarks *135*
References *136*

7 Regulation of Two Electricity Generators with Dead Zone *139*
7.1 Introduction *139*
7.2 Mathematical Model and Regulators for the Electricity Generators with Dead Zone *140*

7.3 Sliding-mode Regulator with Sine Function *147*
7.4 Simulations *149*
7.4.1 Electricity Generators with Dead Zone with Two Static Magnets and Two Dynamic Magnets *150*
7.4.2 Electricity Generators with Dead Zone with Four Static Magnets and Four Dynamic Magnets *153*
7.5 Concluding Remarks *157*
References *157*

8 A Regulator to Attenuate the Perturbations and to Regulate the Rotatory Inverted Pendulum *161*
8.1 Introduction *161*
8.2 Rotary Inverted Pendulum *162*
8.3 The Problem of Regulation for the Pole Placement Strategy *165*
8.4 The Problem of Perturbation Attenuation *165*
8.5 Stable Regulator Which Attenuate the Perturbations for the Rotatory Inverted Pendulum *166*
8.5.1 Stability Analysis of the Process *167*
8.5.2 Controllability of the Process *167*
8.5.3 Regulator Design *167*
8.6 Simulation *168*
8.7 Concluding Remarks *169*
References *171*

9 Proportional-derivative Regulator with Inverse Dead Zone for Pendulum Processes *173*
9.1 Introduction *173*
9.2 Mathematical Model of the Robotic Arms with Dead Zone Inputs *174*
9.3 Mathematical Model of the Pendulum Processes with Dead Zone Inputs *176*
9.4 Proportional Derivative Regulator with Inverse Dead Zone *177*
9.5 Simulations *179*
9.5.1 Example 1 *179*
9.5.2 Example 2 *182*
9.6 Concluding Remarks *188*
References *188*

10 Sliding-mode Regulator of Robotic Arms with Dead Zone *191*
10.1 Introduction *191*
10.2 Mathematical Model of Robotic Arms with Dead Zone and Gravity *192*

10.3 Sliding-mode Regulator for the Regulation of Robotic Arms with Dead Zone and Gravity *195*
10.4 Comparisons *197*
10.4.1 Transelevator *198*
10.4.2 Articulated Robotic Arm *202*
10.5 Concluding Remarks *207*
References *207*

Index *211*

Preface

The goal of this book is to provide a detailed presentation of the estimation and attenuation of perturbations. The estimation and attenuation of perturbations is a very interesting investigation field, since there are many applications of estimation or attenuation of perturbations in mechatronics, energy, and robotic processes.

Perturbations are undesired signals presented as inputs of a process, which affect the outputs. This problem has been studied in many processes. The mechatronic processes of this book are the mechanical process, the thermal process, the three phase electric circuit, the plotter, the suspension process, and the hydraulic process. The energy processes of this book are the wind turbine, the electric vehicle, the gas turbine, the gasification process, and the electricity generators with magnets. The robotic processes of this book are the rotary inverted pendulum, the inverted car pendulum, the Furuta pendulum, the transelevator, and the articulated robotic arm.

This book is expected to be used primarily by researchers and secondarily by students in the areas of regulation, robotics, energy, biological, mechanical, mechatronics, and computing processes.

The recommended use of the book is to be focused on each kind of process because they are different from each other. Chapters 1–4 cover the mechatronic processes, Chapters 5–7 cover the energy processes, and Chapters 8–10 cover the robotic processes.

Mexico City, Mexico
2025

Jose de Jesus Rubio

Acknowledgments

The author thanks the Institution of Engineering and Technology for permitting the reuse portions of works [1, 2] in Chapters 1 and 10; Elsevier for permitting the reuse portions of works [3–5] in Chapters 2, 3, and 5; the Institute of Electrical and Electronics Engineers for permitting the reuse portions of works of [6, 7] in Chapters 4 and 6; the Hindawi Publishing Corporation for permitting the reuse portions of works of [8, 9] in Chapters 7 and 9; and the Sociedad Mexicana de Física to let the reuse portions of works of [10] in Chapter 8 of this book. The author also thanks the Instituto Politécnico Nacional, Secretaría de Ciencia, Humanidades, Tecnología e Innovación, Secretaría de Investigación y Posgrado, and Comisión de Operación y Fomento de Actividades Académicas for their help in this study.

References

1 Rubio J. J., Figueroa M., Perez-Cruz J. H., Bejarano F. J. (2012), Geometric approach and structure at infinite controls for the disturbance rejection, IET Control Theory and Applications, 6 (16), 2528–2537.
2 Rubio J. J. (2017), Sliding mode control of robotic arms with deadzone, IET Control Theory and Applications, 11 (8), 1214–1221.
3 Rubio J. J. (2016), Structure control for the disturbance rejection in two electromechanical processes, Journal of the Franklin Institute, 353 (14), 3610–3631.
4 Rubio J. J. (2016), Hybrid controller with observer for the estimation and rejection of disturbances, ISA Transactions, 65, 445–455.
5 Rubio J. J., Ochoa G., Balcazar R., Pacheco J., (2015), Uniform stable observer for the disturbance estimation in two renewable energy systems, ISA Transactions, 58, 155–164.

6 Rubio J. J., Soriano E., Juarez C. F., Pacheco J., (2017), Sliding mode regulator for the perturbations attenuation in two tank plants, IEEE Access, 5 (1), 20504–20511.

7 Rubio J. J., Martinez D. I., Garcia V., Gutierrez G. J., Vargas T. M., Ochoa G., Balcazar R., Pacheco J., Meda Campaña J. A., Mujica Vargas D., (2020), The perturbations estimation in two gas plants, IEEE Access, 8 (1), 83081–83091.

8 Martinez D. I., Rubio J. J., Aguilar A., Pacheco J., Gutierrez G. J., Garcia V., Vargas T. M., Ochoa G., Cruz D. R., Juarez C. F., (2020), Stabilization of two electricity generators, Complexity, 2020, 1–13.

9 Rubio J. J., Zamudio Z., Pacheco J., Mujica Vargas D., (2013), Proportional derivative control with inverse dead-zone for pendulum systems, Mathematical Problems in Engineering, 2013, 1–9.

10 Rubio J. J., Figueroa M., Perez Cruz J. H., Rumbo J., (2012), Control to stabilize and mitigate disturbances in a rotary inverted pendulum, Revista Mexicana de Fisica E, 58 (2), 107–112.

Introduction

Mathematical models are essential for understanding the process dynamics in open-loop (manual mode) or in closed-loop (automatic) regulator. It is of importance because one mistake in experiments on prototypes could cause accidents or damage to prototypes or devices, while one mistake in the simulation of mathematical models could cause nothing. It helps in the method designs by considering mathematical models as a first stage of design before the experiments. Each process has a different mathematical model; they are mainly classified into two kinds of mathematical models. The first kind of mathematical model is mainly obtained by the Euler–Lagrange method and described by differential equations; they are either derived from data (empirical) and from more fundamental relationships (first principles, physics-based) that rely on knowledge of the process. A combination of the two approaches is often used in practice where fundamental balance equations are adjusted to fit process data. The second kind of mathematical model is mainly obtained by the neural networks and described by difference equations; they are either derived from data (empirical) or from more fundamental relationships (their structure, their learning) that rely on knowledge of the process, a combination of the two approaches is often used in practice where unknown or uncertain parameters are adjusted to fit process data. This book is focused on the first kind of mathematical models, which are mainly obtained by the Euler–Lagrange method.

Perturbations are undesired signals presented as inputs of a process that affect the outputs. Estimation and attenuation of perturbations in the process output are required because they can affect the sensors, actuators, processes, or regulators, causing accidents or unnecessary costs. Additionally, the dead zone can be seen as perturbations in many industrial processes, which can severely limit the process performance. The perturbations study has been drawing much interest in the regulation community because this problem has many recent applications, such as energy, power, aerospace, ocean, mechanical, medical, vehicle, and robotic processes. The applications are detailed as follows.

The perturbation attenuation regulator has been applied to some energy processes. In [1], a robust active perturbation attenuation regulator is introduced for thermoacoustic instability. In [2], a novel active perturbation attenuation regulator is proposed for a proton exchange membrane fuel cell. In [3], a novel model-dependent extended state observer-based active perturbation attenuation regulator is discussed for the pitch regulation of floating offshore wind turbines. In [4], a novel double-loop compensated active perturbation attenuation regulator is discussed for the safe flight of rotorcraft. In [5], a phase-leading cascaded error-based active perturbation attenuation regulator is proposed for a wind turbine. In [6], the active perturbation attenuation regulator is designed to regulate the vibration isolation platform. In [7], a decomposition prediction fractional-order active perturbation attenuation regulator is discussed to regulate each generator of integrated energy processes. In [8], an optimized simplified nonlinear active perturbation attenuation regulator is discussed for a doubly fed induction generator-based wind energy process. In [9], an active perturbation attenuation regulator is presented for microgrids in renewable energy generation. In [10], an adaptive perturbation observer-based nonlinear tracking regulator is developed for a mini excavator proton exchange membrane fuel cell.

The perturbation attenuation regulator has been applied to some power processes. In [11], a phase lead error-based active perturbation attenuation regulator is addressed for an ultra-supercritical unit. In [12], a linear active perturbation attenuation regulator is presented for buck and boost DC/DC der-interfacing converters. In [13], an active perturbation attenuation regulator is proposed for the pressure regulation in a pressurized water reactor nuclear power process. In [14], a linear active perturbation attenuation regulator is addressed for partially shaded photovoltaic-powered battery charging. In [15], a sliding mode active perturbation attenuation regulator is proposed for a DC active power filter. In [16], an enhanced active perturbation attenuation regulator is considered for a photovoltaic grid-connected four-leg inverter. In [17], a modified perturbation attenuation regulator is addressed for a dual active bridge converter. In [18], an enhanced load perturbation attenuation regulator is introduced for a boost converter. In [19], a proportional–integral–derivative regulator for a nonideal buck converter.

The perturbation attenuation regulator has been applied to some aerospace processes. In [20], a dual-loop linear active perturbation attenuation regulator is proposed for a parafoil airdrop process. In [21], multi-input multi-output modified active perturbation attenuation regulator for flow-induced vibration of the wind tunnel test models. In [22], the output feedback active perturbation attenuation regulator is investigated for a valve-regulation cylinder electro-hydraulic servo process. In [23], an active perturbation attenuation regulator is studied for a rigid-liquid-flexible coupling spacecraft. In [24], a novel active perturbation attenuation regulator is suggested for spacecraft relative pose synchronization. In [25], an adaptive non-singular fast terminal sliding mode regulator is proposed for a rotorcraft.

The perturbation attenuation regulator has been applied to some ocean processes. In [26], an innovative composite regulator scheme is proposed for the vessel motion under random wave excitations. In [27], an active perturbation attenuation regulator is proposed for a fully submerged hydrofoil craft. In [28], a constrained underactuated autonomous underwater vehicles method is proposed based on active perturbation attenuation regulator.

The perturbation attenuation regulator has been applied to some mechanical processes. In [29], an active perturbation attenuation regulator is utilized for an automotive suspension process. In [30], an active perturbation attenuation regulator is developed for a permanent magnet synchronous motor. In [31], a maximum power point voltage regulator is designed for an adaptive reference voltage generator.

The perturbation attenuation regulator has been applied to some medical processes. In [32], a linear active perturbation attenuation regulator for epilepsy modulation. In [33], a linear active perturbation attenuation regulator is developed for measuring ultra-weak magnetic fields based on cell rotation. In [34], a physical perturbation attenuation regulator is suggested for limb prostheses.

The perturbation attenuation regulator has been applied to some vehicle processes. In [35], a proportional–derivative regulator is designed for an autonomous vehicle. In [36], an enhanced car-following algorithm utilizing active perturbation attenuation regulator is proposed.

The perturbation attenuation regulator has been applied to some robotic processes. In [37], a linear active perturbation attenuation control is presented for an omnidirectional robot. In [38], two model-free regulators are discussed for continuum robots.

From the aforementioned research, most of the investigations consider the active perturbation attenuation regulator as the main strategy for the perturbation attenuation. So, it would be interesting to study alternative regulation strategies for the perturbation attenuation, and also it would be interesting to study observer strategies for the perturbations estimation in the theory and applications.

The first benefit of this book is the attenuation of perturbations; it is solved by using the following strategies: the geometric approach, the structure at infinity, the structure regulator, the proportional–derivative regulator, the sine function, and the sliding mode regulator. The second benefit of this book is the estimation of perturbations; it is solved by using the following strategies: the linear observer, the nonlinear observer, and the sliding mode observer. In most of the regulators and observers, the stability is warranted by using the Lyapunov strategy.

A perturbation regulator is a process that maintains a desired value by measuring and comparing the current value to the desired value. The difference between the two values is then used to reduce the difference. The goal of perturbation regulator is to reduce inefficiencies that can occur when humans are involved. The geometric regulator is based on the differential geometry and could be used to attenuate perturbations in linear processes. The structure at infinity regulator is a way to characterize the solvability of regulation problems and could be used to attenuate perturbations in linear processes. The structure regulator is an improved version of the structure at infinity regulator and could be used to attenuate perturbations in linear processes. The proportional–derivative regulator is a combination of feedforward and feedback regulator because it operates on both the current process conditions and predicted process conditions and could be used to attenuate perturbations in nonlinear processes. The sine function regulator is an improved version of the proportional–derivative regulator and could be used to attenuate perturbations in nonlinear processes. The sliding mode regulator has a sliding surface and could be used to attenuate perturbations in nonlinear processes. A perturbation observer estimates the uncertainties, perturbations, and dynamics of a process. It can be used to improve regulation performance and tracking. A linear observer has a linear structure and could be used to estimate perturbations of a linear process. A nonlinear observer has a nonlinear structure and uses nonlinear dynamics to estimate the perturbations of a nonlinear mathematical model. A sliding mode observer has a sliding surface and could be used to estimate perturbations of a nonlinear process.

This book contains three parts: Part 1 of Chapters 1–4 contains the estimation and attenuation of perturbations in mechatronic processes, Part 2 of Chapters 5–7 contains the estimation and attenuation of perturbations in energy processes, and Part 3 of Chapters 8–10 contains the estimation and attenuation of perturbations in robotic processes. The mechatronic processes of this book are the mechanical process, the thermal process, the three-phase electric circuit, the plotter, the suspension process, and the hydraulic process. The energy processes of this book are the wind turbine, the electric vehicle, the gas turbine, the gasification process, and the electricity generators with magnets. The robotic processes of this book are the rotary inverted pendulum, the inverted-car pendulum, the Furuta pendulum, the transelevator, and the articulated robotic arm.

The detailed description of chapters in this book is as follows:

In Chapter 1, two strategies are recommended to solve the perturbation attenuation problem. The two strategies are the structure at infinity regulators and the geometric method regulator. The main benefit is that even when the two strategies to solve the perturbation attenuation problem are different, in both cases, if the problem has a solution, there exists an equality in the structure of the regulator functions such that the transfer function of the outputs and the perturbations is null. Both strategies are applied to attenuate the perturbations in a mechanical process and a thermal process. This work is also published in [39].

In Chapter 2, the structure regulator is recommended for the perturbation attenuation. The gains of the recommended strategy are selected to satisfy three objectives: (1) the perturbation attenuation, (2) stability, and (3) convergence of the errors. The structure regulator design is based on the structure at infinity strategy, but the recommended strategy has two benefits: the perturbations are unknown because they are not utilized in the regulator law, and the condition of having equal values in the input and perturbation transfer functions is not required. The recommended strategy is applied for the perturbation attenuation in two processes. This work is also published in [40].

In Chapter 3, a hybrid regulator with observer is recommended for the estimation and attenuation of perturbations. It is based on the combination of the sliding mode strategy and the output feedback strategy. It is divided into two designs: (1) the observer and (2) the regulator with observer. The observer is selected to reach two benefits: (1) to warrant its stability and (2) for the estimation of perturbations. The regulator with observer is selected to reach three benefits: (1) to warrant its stability, (2) for the attenuation of perturbations, and (3) for the decreasing of chattering in the sliding mode behavior. The recommended strategy is applied for the estimation and attenuation of the perturbations in a plotter and a suspension process. This work is also published in [41].

In Chapter 4, a sliding mode regulator is recommended for the perturbation attenuation in the process output and states. The benefit of the recommended regulator is that it does not employ an observer for the perturbations estimation; so, it is more compact. The asymptotic stability of the regulator is warranted via the Lyapunov strategy. The mentioned regulator is employed in two tank processes. This work is also published in [42].

In Chapter 5, an observer for the state and perturbation estimation in two energy processes is recommended. The restrictions of the gains in the recommended observer are found to warrant its stability and the convergence of its error; in addition, these results are utilized to obtain a good estimation. The recommended strategy is applied for the state and perturbation estimation in a wind turbine and an electric vehicle. The wind turbine has a rotatory tower to catch the incoming air to be transformed into electricity, and the electric vehicle has generators connected with its wheels to catch the vehicle's movement to be transformed into electricity. This work is also published in [43].

In Chapter 6, an estimator is recommended for the variables and perturbation estimation in nonlinear processes. The saturation function is utilized in our estimator instead of the signum function to decrease the chattering, and we warrant the estimator convergence by the Lyapunov strategy. The conditions required by our estimator gains are found to reach the variable error convergence, and these gains are utilized for the perturbation estimation. An algorithm is recommended to choose the gains for achieving a satisfactory performance in our estimator. The recommended estimator is applied for the variable and perturbation estimation in the gas turbine and gasification processes. This work is also published in [44].

In Chapter 7, a sliding mode regulator with sine function is recommended for the regulation of electricity generators being affected by magnet interaction nonlinearities and generator perturbations. To reach this goal, our recommended regulator has the following benefits: (1) it starts from the sliding mode regulator with the modifications that the saturation function is utilized to reach a smoother performance instead of the signum function, and the sine function is applied to reach an upper bound in the proportional gain error, (2) it is utilized to reach some chosen constant behaviors for the angle position, angle speed, and current in the electricity generators with dead zone, and (3) its regulation is warranted based on the Lyapunov approach. We show the simulation of the recommended regulator in two electricity generators with dead zone. This work is also published in [45].

In Chapter 8, the combination of two strategies is recommended to regulate and to attenuate the perturbations of a rotatory inverted pendulum. (1) The perturbation is attenuated if there exists a regulator function such that the transfer function of the output and perturbation is null. (2) The pole placement regulator makes the equality of two polynomials: the first polynomial is the regulation process and the second is the desired polynomial. It forces the process to follow the behavior of the desired polynomial, which is stable. This work is also published in [46].

In Chapter 9, a proportional–derivative regulator with inverse dead zone is recommended for the regulation of pendulum processes. The recommended strategy has the benefit that the inverse dead zone is canceled with the pendulum dead zone. Asymptotic stability of the recommended strategy is warranted by the Lyapunov analysis. Simulations of two pendulum processes show the effectiveness of the recommended strategy. This work is also published in [47].

In Chapter 10, a sliding mode regulator is recommended for the regulation of robotic arms with unknown behaviors in the dead zone and gravity. The sliding mode strategy is employed to compensate the unknown behaviors. In addition, the asymptotic stability of the regulator is warranted by the Lyapunov analysis. The application in two robotic arms shows the effectiveness of the recommended strategy. This work is also published in [48].

Mexico City, Mexico
2025 *Jose de Jesus Rubio*

References

1 Liu S., Wang Z., Yang D., Wang Y., Wang Y., Li D., Zhu M., (2025), Robust frequency response-based active disturbance rejection control to mitigate thermoacoustic instability in the Rijke tube burner, Applied Energy, 378, 124779.

2 Zhu C., Li B., Qin Y., Gao M., Liu G., (2025), Temperature management of liquid-cooled fuel cells based on active disturbance rejection control, Applied Thermal Engineering, 258, 124806.

3 Zhao Y., Yang X., Zhao Z., (2025), Pitch control for floating offshore wind turbines via model-dependent extended state observer-based active disturbance rejection control, Ocean Engineering, 316, 119883.

4 Su Z., Shi W., Duan J., Xu L., Zhou M., (2025), Double-loop compensated active disturbance rejection control of electromechanical servo system based on composite disturbance observer, Expert Systems With Applications, 266, 126155.

5 Li X., Wang W., Fang F., Liu J., Chen Z., (2025), Improving active power regulation for wind turbine by phase leading cascaded error-based active disturbance rejection control and multi-objective optimization, Renewable Energy, 243, 122629.

6 Ma C., Li W., Li Q., Liu L., (2025), Fault tolerant control of vibration isolation platform based on fuzzy active disturbance rejection controller, Aerospace Science and Technology, 159, 109945.

7 Lu Q., Huang W., Yin L., (2025), Decomposition prediction fractional-order active disturbance rejection control deep Q network for generation control of integrated energy systems, Applied Energy, 377, 124773.

8 Sobhy A., Elfar M. H., Refaat A., Fawzi M., (2025), Performance evaluation of an optimized simplified nonlinear active disturbance rejection controller for rotor current control of DFIG-based wind energy system, Cleaner Engineering and Technology, 24, 100896.

9 Khamies M., Sayed K., Alrumayh O., Almutairi A., Mahmoud A. A., (2025), Advanced active disturbance rejection control for enhancing frequency stability in low-inertia power grids linked with virtual inertia applications, Heliyon, 11, e42556.

10 Phan V. D., Trinh H.-A., Ahn K. K., (2025), Adaptive disturbance observer-based terminal sliding mode algorithm for a mini excavator proton exchange membrane fuel cell air feeding system, Applied Energy, 382, 125304.

11 Pang D., Niu Y., Du M., (2025), Phase lead error-based active disturbance rejection control for 1000MW ultra-supercritical unit under flexible operation, Energy, 319, 134946.

12 Korompili A., Erin O., Stevic M., Hagenmeyer V., Monti A., (2025), Linear active disturbance rejection control-based voltage controller for buck and boost DC/DC converters in DC distribution grids, IEEE Access, 13, 19085–19109.

13 Zhang Q., Sun A., Xiao L., Wei X., Sun P., (2025), Active disturbance rejection control of pressurizer pressure during startup process in a pressurized water reactor nuclear power plant, Nuclear Engineering and Technology, 57, 103325.

14 Ibrahim A.-W., Xu J., Al-Shamma A. A., Hussein Farh H. M., Dagal I., (2025), Intelligent adaptive PSO and linear active disturbance rejection control: a novel reinitialization strategy for partially shaded photovoltaic-powered battery charging, Computers and Electrical Engineering, 123, 110037.

15 Li G., Wang H., Liu X., Chen J., (2025), A sliding mode active disturbance rejection control of DC active power filter for enhancing DC microgrid voltage robustness, Electric Power Systems Research, 241, 111388.

16 Ali C., Mabrouk D., Ihammouchen S., Bizon N., Habib B., Abdelhalim K., Toufik R., Fouad B. M., (2025), An enhanced active disturbance rejection control scheme for DC voltage regulation in photovoltaic grid-connected four-leg inverter using a sliding mode observer, IEEE Access, 13, 18623–18643.

17 Pal P., Behera R. K., Muduli U. R., (2025), Feedforward repetitive approach to disturbance rejection in DAB converters for EV charging, IEEE Transactions on Industrial Electronics. https://doi.org/10.1109/TIE.2024.3519565

18 Somanshu S. R., Anwar M. N., Kumar R., (2025), Enhancing disturbance rejection in boost converter for CPL: a controller design approach with partial pole placement and approximate model matching, ISA Transactions, 156, 389–400.

19 Ersali C., Hekimoglu B., Yilmaz M., (2025), Designing a filtered proportional integral–derivative controller with disturbance rejection for a nonideal buck converter utilizing an upgraded genetic algorithm and pattern search, Advanced Control for Applications: Engineering and Industrial Systems, 7, e70001.

20 Yang J., Tan P., Sun Q., Sun H., Tao J., Fei Z., Chen Z., (2025), Dual-loop linear active disturbance rejection control for parafoil airdrop systems in wind environment, Aerospace Science and Technology, 158, 109910.

21 Li W., Yang Z., Liu K., Zhu X., Wang W., He S., (2025), Piezoelectric active disturbance rejection control for flow-induced vibration of the wind tunnel test models, Mechanical Systems and Signal Processing, 224, 112010.

22 Bai Y., Yao J., Hu J., Feng G., (2025), Output feedback active disturbance rejection control of an electro-hydraulic servo system based on command filter, Chinese Journal of Aeronautics, 38 (2), 103169.

23 Zhang X., Li S., Zhu H., Wang J., (2025), Attitude tracking of rigid-liquid-flexible coupling spacecraft by active disturbance rejection control, Acta Astronautica, 226, 913–927.

24 Li W., Chen J., Yu Y., Gao C., Yuan J., (2025), Active disturbance rejection control for spacecraft relative motion in libration point orbits subject to actuator saturation and time-delays, Advances in Space Research, 75, 2169–2183.

25 Miranda-Moya A., Castañeda H., Wang H., (2025), Disturbance rejection-based adaptive non-singular fast terminal sliding mode control for a quadrotor under severe turbulent wind, ISA Transactions, 157, 248–257.

26 Lin C., Xin B., Wang S., Zhou S., Chen W., Gao F., (2025), Dynamic characteristics analysis and multi-source disturbances rejection control of a 6-PUS parallel stabilization mechanism on floating base, Mechanism and Machine Theory, 205, 105888.

27 Li B., Xu Z., Mao T., Jia Z., Wang L., Shu Y., Liu H., (2025), Study of the double closed-loop active disturbance rejection control strategy for the longitudinal motion of fully submerged hydrofoil craft, Ocean Engineering, 315, 119714.

28 Feng Z., Yao S., (2025), Dynamic event-triggered active disturbance rejection formation control for constrained underactuated AUVs, IEEE/CAA Journal of Automatica Sinica, 12 (2), 460–462.

29 Nguyen T. A., (2025), Active disturbance rejection control for reducing vehicle body displacement based on an extended state observer and dynamic fuzzy techniques, Journal of Low Frequency Noise, Vibration and Active Control, 0 (0), 1–16.

30 Chung Y., Ahn H., Hu M., Park J., You K., (2025), Novel control strategy for robustness of two DOF Smith predictor via active disturbance rejection method, Journal of the Franklin Institute, 362, 107457.

31 Nagi F., Kumaran V., Mansor M., Verayiah R., Mohamed H. B., Permal N., (2025), Maximum power point tracking controller with online adaptive reference voltage generator for disturbance rejection, Renewable Energy, 241, 122277.

32 Wei W., Wang Z., (2025), Closed-loop seizure modulation via extreme learning machine supervisor based sliding mode disturbance rejection control, Neurocomputing, 617, 129026.

33 Zhao X., Tian P., Zhang H., Wu H., Ye X., Zhang L., Zhao F., Sun J., (2025), Rotation measurement system based on linear active disturbance rejection and quasi-proportional resonant control, Measurement, 244, 116306.

34 Paggetti F., Gherardini M., Ianniciello V., Cirelli S., Cipriani C., (2025), Physical disturbance rejection methods in myokinetic control of prosthetic limbs, IEEE Transactions on Biomedical Engineering. https://doi.org/10.1109/TBME.2025.3536854

35 Li L., Song K., Tang G., Xue W., Xie H., Ma J., (2025), Active disturbance rejection path tracking control of vehicles with adaptive observer bandwidth based on Q-learning, Control Engineering Practice, 154, 106137.

36 Gao H., Yang H., Zhang X., Ren X., Liang F., Yan R., Liu Q., Hu M., Zhang F., Gao J., Bao S., Li K., Li D., Wang D., (2025), Longitudinal velocity control of autonomous driving based on extended state observer, CAAI Transactions on Intelligence Technology, 10, 36–46.

37 Ramirez-Neria M., Madonski R., Hernandez-Martinez E. G., Lozada-Castillo N., Fernandez-Anaya G., Luviano-Juarez A., (2025), Robust trajectory tracking for omnidirectional robots by means of anti-peaking linear active disturbance rejection, Robotics and Autonomous Systems, 183, 104842.

38 Yilmaz C. T., Watson C., Morimoto T. K., Krstic M., (2025), Adaptive model-free disturbance rejection for continuum robots, Automatica, 171, 111949.

39 Rubio J. J., Figueroa M., Perez-Cruz J. H., Bejarano F. J., (2012), Geometric approach and structure at infinite controls for the disturbance rejection, IET Control Theory and Applications, 6 (16), 2528–2537.

40 Rubio J. J., (2016), Structure control for the disturbance rejection in two electromechanical processes, Journal of the Franklin Institute, 353 (14), 3610–3631.

41 Rubio J. J., (2016), Hybrid controller with observer for the estimation and rejection of disturbances, ISA Transactions, 65, 445–455.

42 Rubio J. J., Soriano E., Juarez C. F., Pacheco J., (2017), Sliding mode regulator for the perturbations attenuation in two tank plants, IEEE Access, 5 (1), 20504–20511.

43 Rubio J. J., Ochoa G., Balcazar R., Pacheco J., (2015), Uniform stable observer for the disturbance estimation in two renewable energy systems, ISA Transactions, 58, 155–164.

44 Rubio J. J., Martinez D. I., Garcia V., Gutierrez G. J., Vargas T. M., Ochoa G., Balcazar R., Pacheco J., Meda Campaña J. A., Mujica Vargas D., (2020), The perturbations estimation in two gas plants, IEEE Access, 8 (1), 83081–83091.

45 Martinez D. I., Rubio J. J., Aguilar A., Pacheco J., Gutierrez G. J., Garcia V., Vargas T. M., Ochoa G., Cruz D. R., Juarez C. F., (2020), Stabilization of two electricity generators, Complexity, 2020, 1–13.

46 Rubio J. J., Figueroa M., Perez Cruz J. H., Rumbo J., (2012), Control to stabilize and mitigate disturbances in a rotary inverted pendulum, Revista Mexicana de Fisica E, 58 (2), 107–112.

47 Rubio J. J., Zamudio Z., Pacheco J., Mujica Vargas D., (2013), Proportional derivative control with inverse dead-zone for pendulum systems, Mathematical Problems in Engineering, 2013, 1–9.

48 Rubio J. J., (2017), Sliding mode control of robotic arms with deadzone, IET Control Theory and Applications, 11 (8), 1214–1221.

1

Geometric Method and Structure at Infinity Regulators for the Perturbation Attenuation

1.1 Introduction

The perturbation attenuation problem is an important topic of the regulator processes, both in theory and applications. This problem has been studied in many processes. As soon as more perturbations have been found, more strategies have been proposed to attenuate them.

Some examples of the aforementioned facts can be found in [1–6].

Chen et al. [1] investigate the problem of state feedback robust stabilization for discrete-time fuzzy singularly perturbed processes with parameter uncertainty. In [2], a perturbation observer is inserted into the position-loop, which compensates for the velocity error owing to the presence of unparameterized uncertainties. In [3], the attitude regulator of a quadrotor aircraft is recommended subject to a class of perturbations. In [4], regulators are proposed to compensate for the perturbations of the base movements for a robotic manipulator mounted on a moving base. The study of [5] deals with the problem of robust adaptive L_∞-gain neural filter design for a class of uncertain processes with unknown nonlinearities and persistently bounded perturbations. In [6], the authors propose a strategy that emphasizes the implementation issue and employs state feedback integral regulator to achieve nonzero perturbation attenuation and zero steady-state error. So, the perturbation attenuation regulator has become a challenge with increasing demands of the disturbed processes.

This chapter studies two different strategies to attenuate the perturbations: the geometric method regulator [7–9] and the structure at infinity regulator [10–12]. Both strategies will be applied to attenuate the perturbations in a mechanical process and a thermal process.

The rest of the chapter is organized as follows. Section 1.2 shows the main notations of the chapter. The perturbation attenuation problem is introduced in Section 1.3. Section 1.4 presents the geometric method regulator to attenuate the perturbations. The structure at infinity regulator to attenuate the perturbations is described in Section 1.5. Section 1.6 applies both strategies to attenuate the perturbations in a mechanical process. Both strategies are applied to attenuate the perturbations in a thermal process in Section 1.7. Section 1.8 presents the conclusion and the future research.

1.2 Notation

$\mathcal{U}$, $\mathcal{X}$, $\mathcal{Y}$ are linear subspaces with elements ν, χ, y, respectively. A, B, C are the matrices, which represent the linear transformations $A : \mathcal{X} \to \mathcal{X}, B : \mathcal{U} \to \mathcal{X}$, and $C : \mathcal{X} \to \mathcal{Y}$, respectively. $A + B$ are the addition of matrices or vectors. $\mathcal{X} \oplus \mathcal{Y}$ are the addition of subspaces. $\mathcal{X} \cap \mathcal{Y}$ are the intersection of subspaces. $\ker C$ is a subspace named the kernel of C, ImB is a subspace named the image of B, $\mathcal{X} \subset \mathcal{Y}$ means a subspace $\mathcal{X}$ is contained in the subspace $\mathcal{Y}$, $\chi \in \mathcal{X}$ means χ is an element of the subspace $\mathcal{X}$. $\langle A|\mathrm{Im}B\rangle$ is the controllable space of the process.

1.3 Perturbation Attenuation Problem

Consider the following linear process (A, B, C) given as:

$$\begin{aligned} \dot{\chi} &= A\chi + B\nu, \\ y &= C\chi, \end{aligned} \tag{1.1}$$

where $t \geq 0$, $\chi \in \mathcal{X} \subseteq \Re^n$ is the state, $\nu \in \mathcal{U} \subseteq \Re^m$ is the regulator function, and $y \in \mathcal{Y} \subseteq \Re^p$ is the output. A, B, and C are the matrices that represent the linear transformations $A : \mathcal{X} \to \mathcal{X}, B : \mathcal{U} \to \mathcal{X}$, and $C : \mathcal{X} \to \mathcal{Y}$, respectively.

The perturbation process (A, B, C, E) is:

$$\begin{aligned} \dot{\chi} &= A\chi + B\nu + Ed, \\ y &= C\chi, \end{aligned} \tag{1.2}$$

where $t \geq 0$, $d \in \mathcal{D} \subseteq \Re^l$ is the perturbations and E is the matrix that represents the linear transformation $E : \mathcal{D} \to \mathcal{X}$.

There are several variants of the perturbations decoupling problems; the one being recommended in this chapter is usually named the known perturbations decoupling problem.

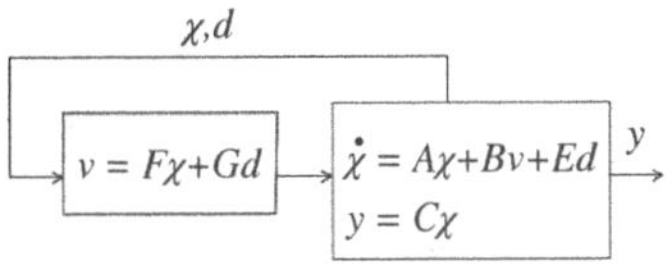

Figure 1.1 Block diagram of the regulator process.

Theorem 1.1: *The perturbation attenuation problem has a solution if there exists a state feedback $F : \mathcal{X} \rightarrow \mathcal{U}$ and a function $G : \mathcal{D} \rightarrow \mathcal{U}$, such that the transfer function of the outputs and the perturbations is null because of the following regulator function:*

$$\nu = F\chi + Gd, \tag{1.3}$$

where the transfer function of the outputs and the perturbations is:

$$G(s) = C(sI - A - BF)^{-1}(BG + E). \tag{1.4}$$

Proof. See [13] for the proof. ■

Figure 1.1 shows the block diagram of the regulator function ν to attenuate the perturbations d in the output y of the process (1.2).

Remark 1.1: Sections 1.4 and 1.5 show two strategies to find F and G, which are substituted in the regulator function (1.3).

Remark 1.2: In both the known perturbations decoupling problem and the unknown perturbations decoupling problem, the benefit is to find a regulator function such that the transfer function of the outputs and the perturbations is null. Nevertheless, in the first problem, the perturbations are known, while in the second problem the perturbations are unknown.

Remark 1.3: The perturbations are not required in the regulator function (1.3) when $G = 0$. So, in this particular case of the general benefit, the known perturbations decoupling problem of this study becomes the unknown perturbations decoupling problem. This fact is recommended in some processes, as is the example of the thermal process (see Section 1.7).

1.4 Perturbation Attenuation Via Geometric Method Regulator

This strategy uses the state equations and invariant subspaces algebra. The benefit is to confine the perturbations to an invariant subspace by a suitable regulator function such that the perturbations do not affect the process outputs.

Definition 1.1: ker(C) is defined as:

$$\ker(C) = \{\chi \in \mathcal{X} | C\chi = 0\}, \tag{1.5}$$

where ker(C) is a subspace of $\mathcal{X} \subseteq \mathbb{R}^n$. dim ker($C$) is given as:

$$\dim \ker(C) = n - \operatorname{rank}(C). \tag{1.6}$$

Definition 1.2: Im(B) is defined as:

$$\operatorname{Im}(B) = \{\chi = B\nu \, | \nu \in \mathcal{U}\}, \tag{1.7}$$

where Im(B) is a subspace of $\mathcal{X} \subseteq \mathbb{R}^n$. dim Im($B$) is given as:

$$\dim \operatorname{Im}(B) = \operatorname{rank}(B). \tag{1.8}$$

Definition 1.3: The inverse image of the subspace $\mathcal{S}$ under A is:

$$A^{-1}\mathcal{S} = \{\chi \in \mathcal{X} | A\chi \in \mathcal{S}\}, \tag{1.9}$$

where $A^{-1}\mathcal{S}$ is a subspace of $\mathcal{X} \subseteq \mathbb{R}^n$. dim $\left(A^{-1}\mathcal{S}\right)$ is given as:

$$\dim\left(A^{-1}\mathcal{S}\right) = \dim(\ker\ A) \oplus \dim\left(\mathcal{S} \cap \operatorname{Im}A\right). \tag{1.10}$$

Definition 1.4: A subspace $\mathcal{V} \subseteq \mathbb{R}^n$ is A-invariant if every trajectory of χ initialized in $\mathcal{V}$ never goes out of $\mathcal{V}$, it is named $A\mathcal{V} \subset \mathcal{V}$ [7].

Definition 1.5: A subspace $\mathcal{V} \subseteq \mathbb{R}^n$ is (A, ImB)-regulator invariant if every trajectory of χ initialized $\mathcal{V}$ can be kept in $\mathcal{V}$ by a suitable regulator function ν, i.e., if it exists $F{:}\mathcal{X} \to \mathcal{U}$ defined as $F\chi = \nu$ such that [7]:

$$(A + BF)\mathcal{V} \subset \mathcal{V}. \tag{1.11}$$

Property 1: If $\mathcal{V}$ and $\mathcal{W}$ are (A, ImB)-regulator invariants, then $\mathcal{V} + \mathcal{W}$ is (A, ImB)-maximal regulator invariant set [7, 9, 14, 15].

Consequence 1: For the subspace $\mathcal{P} \subset \mathcal{X}$, there exists a supreme subspace (A, ImB)-invariant contained in $\mathcal{P}$. This supreme subspace can be obtained as the limit of the following not increasing algorithm [7, 9]:

$$\begin{aligned} \mathcal{P}^0 &= \mathcal{P}, \\ \mathcal{P}^{i+1} &= \mathcal{P} \cap A^{-1}\left(\mathcal{P}^i \oplus \operatorname{Im}B\right). \end{aligned} \tag{1.12}$$

Lemma 1: ([14, 15]) *The subspace* $\mathcal{V} \subset \mathcal{X}$ *is* $(A, \text{Im}B)$*-invariant if and only if:*

$$A\mathcal{V} \subset \mathcal{V} \oplus \text{Im}B, \tag{1.13}$$

Proof. See [13] for the proof. ■

If χ can be measured, then the perturbation attenuation problem has a solution and there exists a regulator function $\nu = F\chi$ such that:

$$\langle A + BF|\text{Im}E\rangle \subset \ker\ C, \tag{1.14}$$

where $\langle A + BF|\text{Im}E\rangle = \text{Im}E \oplus (A + BF)\,\text{Im}E \oplus (A + BF)^2\text{Im}E \oplus \cdots \oplus (A + BF)^{n-1}\text{Im}E$.

Substituting $\nu = F\chi$ in the process (A, B, C, E) gives:

$$\begin{aligned} \dot{\chi} &= (A + BF)\chi + Ed, \\ y &= C\chi, \end{aligned} \tag{1.15}$$

Theorem 1.2: *(Unknown perturbations decoupling problem [14, 15]) The perturbation attenuation problem has a solution if and only if:*

$$\text{Im}E \subset \mathcal{V}^*, \tag{1.16}$$

where $\mathcal{V}^*$ *is the supreme of the family of the subspaces* $(A, \text{Im}B)$*-invariants contained in* $\ker C$.

Proof. See [13] for the proof. ■

Corollary 1.1: *The perturbation attenuation problem has a solution if there exists a regulator function* $\nu = F\chi + Gd$*, such that:*

$$\text{Im}E \subset \mathcal{V}^* \oplus \text{Im}B. \tag{1.17}$$

Now, substituting $\nu = F\chi + Gd$ in the process (A, B, C, E) gives:

$$\begin{aligned} \dot{\chi} &= (A + BF)\chi + (BG + E)d, \\ y &= C\chi. \end{aligned} \tag{1.18}$$

Proposition 1: $\text{Im}(BG + E) \subset \mathcal{V}^*$ *if and only if* $\text{Im}\,E \subset \mathcal{V}^* \oplus \text{Im}\,B$.

Proof. Consider that $\text{Im}(BG + E) \subset \mathcal{V}^*$, then:

$$\begin{aligned} (BG + E)\upsilon_1 &= \upsilon_2 \in \mathcal{V}^*, \\ BG\upsilon_1 + E\upsilon_1 &= \upsilon_2, \\ E\upsilon_1 &= \upsilon_2 - BG\upsilon_1, \end{aligned} \tag{1.19}$$

so,

$$\operatorname{Im} E \subset \mathcal{V}^* \oplus \operatorname{Im} B. \tag{1.20}$$

Introduce $\{e_1, \dots, e_k\}$ as a base of $\mathcal{V}^*$, then $Ee_i = \upsilon_i + b_i \; i = 1, \dots, k$, where $\upsilon_i \in \mathcal{V}^*$, $b_i = B\nu_i$. Define $\mathcal{G} : \mathcal{D} \to \mathcal{U}$ such that $Ge_i = \nu_i$, then:

$$\begin{aligned} Ee_i &= \upsilon_i + b_i \\ &= \upsilon_i + B\nu_i \\ &= \upsilon_i + BGe_i, \end{aligned} \tag{1.21}$$

so,

$$\begin{aligned} &Ee_i - BGe_i = \upsilon_i \\ &\Rightarrow (E - BG)e_i = \upsilon_i \in \mathcal{V}^*, \end{aligned} \tag{1.22}$$

and

$$\operatorname{Im}(E - BG) \subset \mathcal{V}^*. \tag{1.23}$$

■

1.5 Perturbation Attenuation Via Structure at Infinity Regulator

If the structure at infinity of the process and that of the perturbations process without a regulator are equal, then the process lets to attenuate the perturbations. So, a regulator function can be applied to attenuate the perturbations in the process outputs.

Definition 1.6: A rational function is proper when $gr(num) \leq gr(den)$ and is strictly proper when $gr(num) < gr(den)$.

The transfer function of the process (A, B, C) is:

$$H(s) = C(sI - A)^{-1}B. \tag{1.24}$$

The transfer function of the perturbations process without regulator (A, E, C) is:

$$H^d(s) = C(sI - A)^{-1}E. \tag{1.25}$$

The matrices $H(s)$ and $H^d(s)$ are rational and strictly proper.

Definition 1.7: A rational function $f(s) \neq 0$ has a zero at infinity ($s = \infty$) if [16]:

$$f(s) = s^{-r} g(s), \tag{1.26}$$

where (1) r is a positive number, (2) $\lim_{s\to\infty} g(s) < \infty$, and (3) $\lim_{s\to\infty} g(s) \neq 0$. It means that $f(s)$ has a zero at infinity with order r.

Definition 1.8: In the case of a variable s, a function is named a rational function if and only if it can be written as $f(s) = \frac{p(s)}{q(s)}$, where $p(s)$ and $q(s)$ are polynomial functions in s and $q(s)$ is not the zero polynomial. A rational matrix function is an array of rational functions [17].

Theorem 1.3: *There exists (not unique) bi-proper matrices B_1 and B_2 such that [10]:*

$$H(s) = B_1(s) \begin{bmatrix} \Delta(s) & 0 \\ 0 & 0 \end{bmatrix} B_2(s), \tag{1.27}$$

where $\Delta(s) = \text{diag}[s^{-n_1}, \ldots, s^{-n_r}]$. The integers n_r are named the orders of the zeros at infinity, and the list of integers $\{n_1, \ldots, n_r\}$ is the structure at infinity and is named $\sum_\infty H(s)$.

The Theorem 1.4 gives the structure at infinity strategy to obtain F and G to be substituted in the perturbation attenuation regulator function (1.3).

Theorem 1.4: *The following propositions are equivalent ([10], [11]):*

1. The perturbation attenuation problem has a solution if the structure at infinity $H(s)$ of (1.24) and $H^d(s)$ of (1.25) are equal, it is named:

$$\sum_{\infty} [H(s)\; H^d(s)] = \sum_{\infty} [H(s)\; 0]. \tag{1.28}$$

2. The perturbations are attenuated by the following regulator function:

$$\nu = F\chi + Gd, \tag{1.29}$$

such that:

$$C(sI - A - BF)^{-1}(BG + E) = 0. \tag{1.30}$$

Proof. The proof is presented in [10, 11]. ■

Remark 1.4: If the perturbation attenuation problem has a solution (1.28), solving (1.30), gives F and G, which are substituted in the regulator function (1.29) to attenuate the perturbations in the process (1.2).

1.6 Mechanical Process

Figure 1.2 shows the mechanical process. m_1, m_2 are the masses of the blocks 1 and 2 (kg), m_3 is the mass of the block 3 (kg), f_1, f_2 are the forces of the input of the press (N), d_1, d_2 are the undesired forces (N), z_1, z_2, z_3 are the movements of the blocks (m), k is the elastic constant (N/m), and b is the shock absorber coefficient (Ns/m). The forces are applied in the blocks 1 and 2; these blocks act over the block 3.

1.6.1 The States-space Process

To obtain the mathematical model of the mechanical process, consider:

$$\begin{aligned} m_1\ddot{z}_1 &= f_1 + d_1, \\ m_2\ddot{z}_2 &= f_2 + d_2, \\ m_1\ddot{z}_3 &= f_1 + d_1 + f_2 + d_2 - kz_3 - b\dot{z}_3. \end{aligned} \tag{1.31}$$

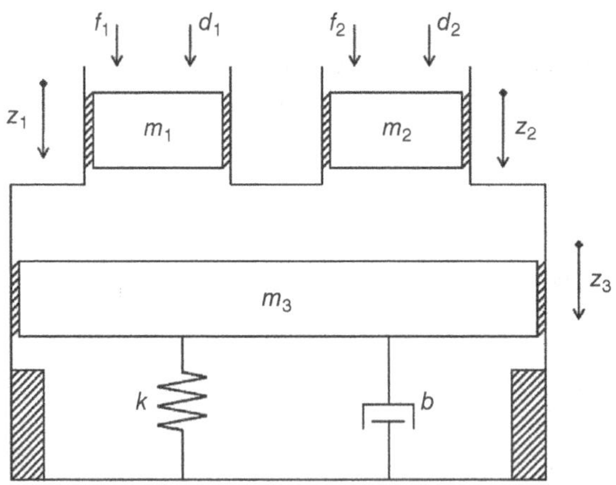

Figure 1.2 Mechanical process and its parameters.

Define the states as $\chi_1 = z_1$, $\chi_2 = \dot{z}_1$, $\chi_3 = z_2$, $\chi_4 = \dot{z}_2$, $\chi_5 = z_3$, $\chi_6 = \dot{z}_3$, the inputs as $\nu_1 = f_1$, $\nu_2 = f_2$, and the outputs as $y_1 = \chi_1$, $y_2 = \chi_3$, $y_3 = \chi_5$; so, the following state-space process is obtained:

$$
\begin{aligned}
\dot{\chi}_1 &= \chi_2, \\
\dot{\chi}_2 &= \frac{1}{m_1}\nu_1 + \frac{1}{m_1}d_1, \\
\dot{\chi}_3 &= \chi_4, \\
\dot{\chi}_4 &= \frac{1}{m_2}\nu_2 + \frac{1}{m_2}d_2, \\
\dot{\chi}_5 &= \chi_6, \\
\dot{\chi}_6 &= \frac{1}{m_3}(\nu_1 + \nu_2) + \frac{1}{m_3}(d_1 + d_2) - \frac{k}{m_3}\chi_5 - \frac{b}{m_3}\chi_6, \\
y_1 &= \chi_1, \\
y_2 &= \chi_3, \\
y_3 &= \chi_5.
\end{aligned}
\tag{1.32}
$$

If $m_1 = m_2 = 50$ kg, $m_3 = 200$ kg, $k = 8000$ N/m, $b = 50$ N s/m, then the matrices of the process (A, B, C) are:

$$
A = \begin{bmatrix}
0 & 1 & 0 & 0 & 0 & 0 \\
0 & 0 & 0 & 0 & 0 & 0 \\
0 & 0 & 0 & 1 & 0 & 0 \\
0 & 0 & 0 & 0 & 0 & 0 \\
0 & 0 & 0 & 0 & 0 & 1 \\
0 & 0 & 0 & 0 & -\frac{8000}{200} & -\frac{50}{200}
\end{bmatrix},
\tag{1.33}
$$

$$
B = \begin{bmatrix}
0 & 0 \\
\frac{1}{50} & 0 \\
0 & 0 \\
0 & \frac{1}{50} \\
0 & 0 \\
\frac{1}{200} & \frac{1}{200}
\end{bmatrix},
\tag{1.34}
$$

$$C = \begin{bmatrix} 1 & 0 & 0 & 0 & 0 & 0 \\ 0 & 0 & 1 & 0 & 0 & 0 \\ 0 & 0 & 0 & 0 & 1 & 0 \end{bmatrix}, \tag{1.35}$$

and the matrix of the perturbations is:

$$E = \begin{pmatrix} 0 & 0 \\ \frac{1}{50} & 0 \\ 0 & 0 \\ 0 & \frac{1}{50} \\ 0 & 0 \\ \frac{1}{200} & \frac{1}{200} \end{pmatrix}. \tag{1.36}$$

1.6.2 Perturbation Attenuation Via the Structure at Infinity Regulator

Obtaining the transfer function $H(s)$:

$$H(s) = \begin{bmatrix} \frac{1}{50s^2} & 0 \\ 0 & \frac{1}{50s^2} \\ \frac{1}{50\,(s + 4s^2 + 160)} & \frac{1}{50\,(s + 4s^2 + 160)} \end{bmatrix}. \tag{1.37}$$

Obtaining the transfer function $H^d(s)$:

$$H^d(s) = \begin{bmatrix} \frac{1}{50s^2} & 0 \\ 0 & \frac{1}{50s^2} \\ \frac{1}{50\,(s + 4s^2 + 160)} & \frac{1}{50\,(s + 4s^2 + 160)} \end{bmatrix}. \tag{1.38}$$

As $H(s)$ of (1.37) and $H^d(s)$ of (1.38) are equal, the structure at infinity of $H(s)$ and $H^d(s)$ are equal.

As $\sum_\infty \left[H(s)\, H^d(s)\right] = \sum_\infty \left[H(s)\, 0\right]$, then the perturbations can be attenuated in the process, i.e., there exists a regulator function $v = F\chi + Gd$, where:

$$F = \begin{bmatrix} f_{11} & f_{12} & f_{13} & f_{14} & f_{15} & f_{16} \\ f_{21} & f_{22} & f_{23} & f_{24} & f_{25} & f_{26} \end{bmatrix}, \tag{1.39}$$

and

$$G = g_{11}. \tag{1.40}$$

Substituting all the data in the following equation (see Theorem 1.4):

$$C(sI - A - BF)^{-1}(BG + E) = 0, \tag{1.41}$$

So,

$$F = \begin{bmatrix} 1 & 1 & 1 & 1 & 1 & 1 \\ -1 & -1 & -1 & -1 & -1 & -1 \end{bmatrix}, \tag{1.42}$$

and

$$G = -1. \tag{1.43}$$

So, the regulator function turns out to be:

$$\nu = \begin{bmatrix} 1 & 1 & 1 & 1 & 1 & 1 \\ -1 & -1 & -1 & -1 & -1 & -1 \end{bmatrix} \chi - d. \tag{1.44}$$

1.6.3 Perturbation Attenuation Via the Geometric Method Regulator

The invariant subspace $\mathcal{V}^*$ is calculated in this strategy. Start with ker C:

$$C\chi = \begin{bmatrix} 1 & 0 & 0 & 0 & 0 & 0 \\ 0 & 0 & 1 & 0 & 0 & 0 \\ 0 & 0 & 0 & 0 & 1 & 0 \end{bmatrix} \begin{bmatrix} \chi_1 \\ \chi_2 \\ \chi_3 \\ \chi_4 \\ \chi_5 \\ \chi_6 \end{bmatrix} = \begin{bmatrix} \chi_1 \\ \chi_3 \\ \chi_5 \end{bmatrix} = \begin{bmatrix} 0 \\ 0 \\ 0 \end{bmatrix}. \tag{1.45}$$

Equation (1.45) implies that every vector in $\ker C$ should have coordinates satisfying:

$$\chi_1 = 0, \chi_3 = 0, \chi_5 = 0. \tag{1.46}$$

Then,

$$\begin{bmatrix} \chi_1 & \chi_2 & \chi_3 & \chi_4 & \chi_5 \end{bmatrix}^T = p_2\chi_2 + p_4\chi_4 + p_6\chi_6, \tag{1.47}$$

where $p_2 = \begin{bmatrix} 0 & 1 & 0 & 0 & 0 & 0 \end{bmatrix}^T$, $p_4 = \begin{bmatrix} 0 & 0 & 0 & 1 & 0 & 0 \end{bmatrix}^T$, and $p_6 = \begin{bmatrix} 0 & 0 & 0 & 0 & 0 & 1 \end{bmatrix}^T$. So:

$$\mathcal{V}^0 = \text{gen}\{p_2, p_4, p_6\} = \ker\ C. \tag{1.48}$$

As dim(ImB) = rank(B) = 2, then,

$$\text{Im}(B) = \text{gen}\{b_1, b_2\}, \tag{1.49}$$

where $b_1 = \begin{bmatrix} 0 & 4 & 0 & 0 & 0 & 1 \end{bmatrix}^T$ and $s_3 = \begin{bmatrix} 0 & 0 & 0 & 4 & 0 & 1 \end{bmatrix}^T$. So,

$$\mathcal{S}_1 = \text{Im}\ B \oplus \mathcal{V}^0 = \text{gen}\{p_2, p_4, p_6\}. \tag{1.50}$$

As $\dim\left(A^{-1}\mathcal{S}\right) = \dim(\ker A) \oplus \dim\left(\mathcal{S} \cap \text{Im}A\right) = 2 + 1 = 3$ and obtaining $A^{-1}\mathcal{S}$ gives:

$$A \begin{bmatrix} \chi_1 \\ \chi_2 \\ \chi_3 \\ \chi_4 \\ \chi_5 \\ \chi_6 \end{bmatrix} = \begin{bmatrix} \chi_2 \\ 0 \\ \chi_4 \\ 0 \\ \chi_6 \\ -40\chi_5 - \frac{1}{4}\chi_6 \end{bmatrix} \in \mathcal{S}_1, \tag{1.51}$$

so,

$$A^{-1}\mathcal{S} = \text{gen}\{s_1, s_3, s_5\}, \tag{1.52}$$

where $s_1 = \begin{bmatrix} 1 & 0 & 0 & 0 & 0 & 0 \end{bmatrix}^T$, $s_3 = \begin{bmatrix} 0 & 0 & 1 & 0 & 0 & 0 \end{bmatrix}^T$, and $s_5 = \begin{bmatrix} 0 & 0 & 0 & 0 & -\frac{1}{40} & 0 \end{bmatrix}^T$. So,

$$\mathcal{V}^1 = \ker\ C \cap A^{-1}\mathcal{S}_1 = \text{gen}\{0\}. \tag{1.53}$$

Making a similar procedure gives:

$$\mathcal{V}^2 = \ker\ C \cap A^{-1}\mathcal{S}_2 = \text{gen}\{0\}. \tag{1.54}$$

As $\mathcal{V}^1 = \mathcal{V}^2$, then,

$$\mathcal{V}^* = \text{gen}\{0\}. \tag{1.55}$$

It can be observed that gen $\{e_1, e_2\} = \text{Im}E \subset \mathcal{V}^* \oplus \text{Im}\, B = \text{gen}\, \{b_1, b_2\}$, i.e.,:

$$\begin{aligned} e_1 &\in \mathcal{V}^* + \text{Im}\, B, \\ e_2 &\in \mathcal{V}^* + \text{Im}\, B, \end{aligned} \tag{1.56}$$

where $e_1 = \begin{bmatrix} 0 & 4 & 0 & 0 & 0 & 1 \end{bmatrix}^T$ and $e_2 = \begin{bmatrix} 0 & 0 & 0 & 4 & 0 & 1 \end{bmatrix}^T$.

As $\text{Im}E \subset \mathcal{V}^* \oplus \text{Im}B$, then there exists a regulator function $\nu = F\chi + Gd$, which satisfies $(A + BF)\mathcal{V}^* \subset \mathcal{V}^* \oplus \text{Im}B$. Set:

$$F = \begin{bmatrix} f_{11} & f_{12} & f_{13} & f_{14} & f_{15} & f_{16} \\ f_{21} & f_{22} & f_{23} & f_{24} & f_{25} & f_{26} \end{bmatrix}. \tag{1.57}$$

Then,

$$(A + BF)\mathcal{V}^* = \mathcal{V}^*. \tag{1.58}$$

The constants f_{ij} are found to satisfy (1.58). F is obtained as:

$$F = \begin{bmatrix} 1 & 1 & 1 & 1 & 0 & 0 \\ -1 & -1 & -1 & -1 & 0 & 0 \end{bmatrix}. \tag{1.59}$$

To obtain $G = g_{11}$, substitute the matrix F in following equation:

$$C(sI - A - BF)^{-1}(BG + E) = 0. \tag{1.60}$$

So, G is obtained as:

$$G = -1. \tag{1.61}$$

So, the regulator function turns out to be:

$$\nu = \begin{bmatrix} 0 & 0 & 0 & 0 & 0 & 0 \\ 0 & 0 & 0 & 0 & 0 & 0 \end{bmatrix} \chi - d. \tag{1.62}$$

1.6.4 Simulation

For the simulation of the process, the initial conditions are $\chi_0 = [0\ \ 0\ \ 0\ \ 0\ \ 0\ \ 0]^T$; so, the blocks start in the rest position, and it is desired that the blocks remain at rest. The two undesired perturbations are $d_1 = d_2 = 100$ N, which force the blocks to have a motion when a regulator is not used. Figure 1.3 shows the regulator functions for the mechanical process. The comparisons for the states of the mechanical process are shown in Figure 1.4. Figure 1.5 shows the comparisons for the outputs of the mechanical process.

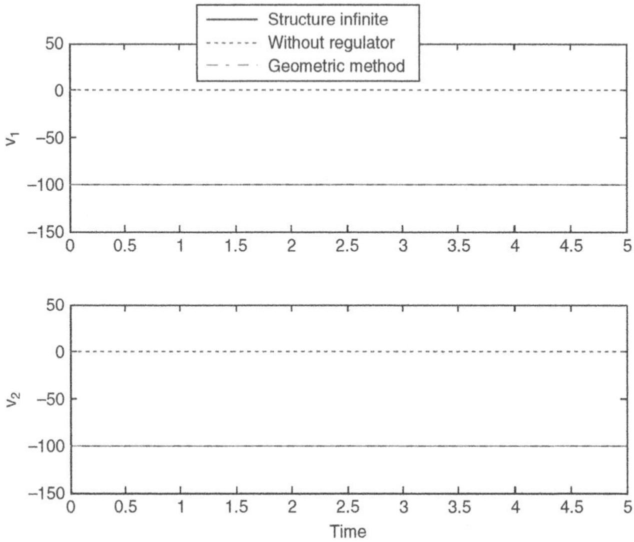

Figure 1.3 Regulator functions for the mechanical process.

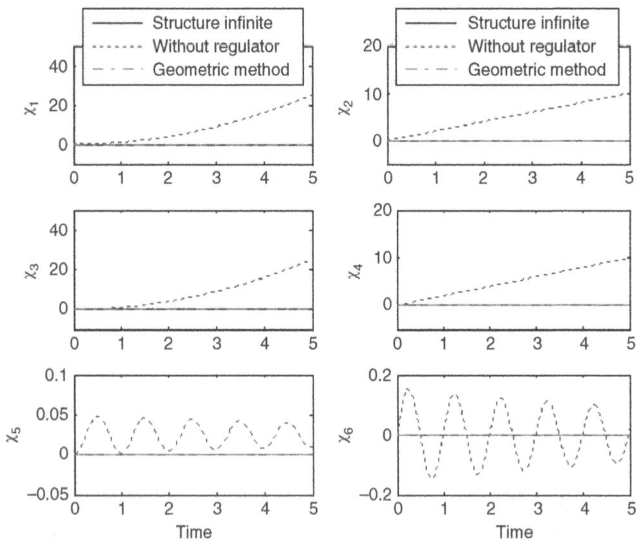

Figure 1.4 States of the mechanical process.

From Figure 1.4, it can be seen that the blocks have a motion when the regulator is not used because of the perturbations, but the blocks do not have a motion when the regulator is used; states χ_5 and χ_5 show the effect of the elastic and the shock absorber. From Figure 1.5, it can be observed that the outputs of the mechanical

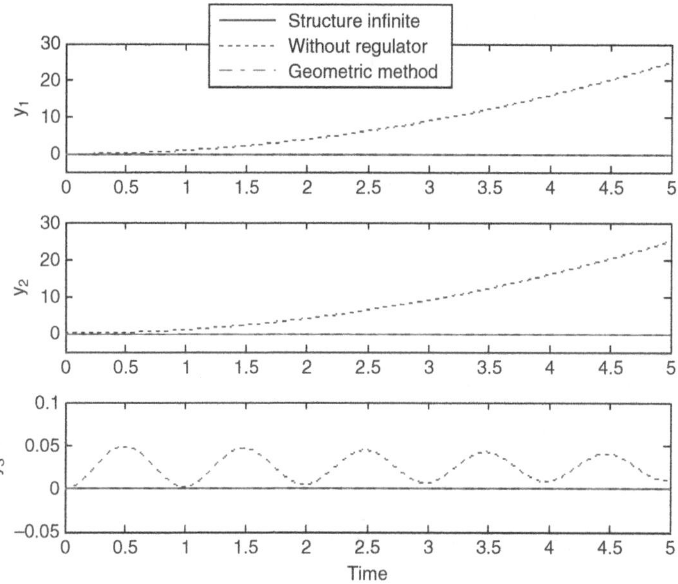

Figure 1.5 Outputs of the mechanical process.

process without a regulator have a motion due to the perturbations, but the outputs of the mechanical process with regulator do not have a motion.

Remark 1.5: From (1.44), (1.62), and Figures 1.3–1.5, it is shown that despite the regulator functions obtained with the two different strategies are different, both regulator functions can attenuate the perturbations in the mechanical process.

1.7 Thermal Process

Figure 1.6 shows the thermal process [14, 15]. t_0 is the temperature of the water that falls in the tanks 1 and 2 (°C), t_1, t_2, and t_3 are the temperatures of the water for the tanks 1, 2, and 3, respectively (°C), f_1 and f_2 are the flows of the water for the tanks 1 and 2 (m^3/s), c_1, c_2, and c_3 are the calorific capacities of the water for the tanks 1, 2, and 3, respectively, ($kcal/m^3$ °C), w is the heating energy used by the tanks 1 and 3 ($kcal/m^3$).

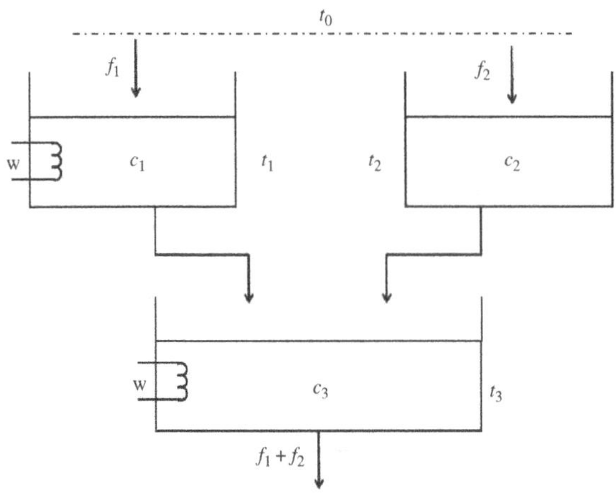

Figure 1.6 Thermal process and its parameters.

The state-space equations of the thermal process are:

$$\begin{aligned}
\dot{t}_1 &= -\frac{f_1}{c_1} t_1 + \frac{1}{C_1} w + \frac{f_1}{c_1} t_0, \\
\dot{t}_2 &= -\frac{f_2}{c_2} t_2 + \frac{f_2}{c_2} t_0, \\
\dot{t}_3 &= \frac{f_1}{c_3} t_1 + \frac{f_2}{c_3} t_2 - \frac{f_1 + f_2}{c_3} t_3 + \frac{1}{c_1} w, \\
y &= t_3,
\end{aligned} \tag{1.63}$$

where t_0 is the perturbations of the process, w is the regulator input of the process, t_1, t_2, and t_3 are the states of the process, and y is the output of the process. If $f_1 = f_2 = 0.5$ m^3/s and $c_1 = c_2 = c_3 = 1000$ kcal/m^3 °C, then the matrices of the process (A, B, C) are:

$$\begin{aligned}
A &= \begin{bmatrix} -\frac{0.5}{1000} & 0 & 0 \\ 0 & -\frac{0.5}{1000} & 0 \\ \frac{0.5}{1000} & \frac{0.5}{1000} & -\frac{1}{1000} \end{bmatrix}, \\
B &= \begin{bmatrix} \frac{1}{1000} \\ 0 \\ \frac{1}{1000} \end{bmatrix}, \\
C &= \begin{bmatrix} 0 & 0 & 1 \end{bmatrix}.
\end{aligned} \tag{1.64}$$

The matrix of the perturbations is:

$$E = \begin{bmatrix} \frac{0.5}{1000} \\ \frac{0.5}{1000} \\ 0 \end{bmatrix}. \tag{1.65}$$

1.7.1 Perturbation Attenuation Via the Structure at Infinity Regulator

Obtaining the transfer function $H(s)$ as follows:

$$\begin{aligned} H(s) &= \frac{1}{3000s + 2 \times 10^6 s^2 + 1} \\ &= s^{-2}\left(\frac{1}{3000s^{-1} + 2 \times 10^6 + 1s^{-2}}\right). \end{aligned} \tag{1.66}$$

As,

$$\lim_{s \to \infty}\left(\frac{1}{3000s^{-1} + 2 \times 10^6 + 1s^{-2}}\right) \neq 0, \tag{1.67}$$

then the order of the structure at infinity of the process is 2. Obtaining the transfer function of the perturbations process $H^d(s) = C(sI - A)^{-1}E$ as follows:

$$\begin{aligned} H^d(s) &= \frac{1}{3000s + 2 \times 10^6 s^2 + 1} \\ &= s^{-2}\left(\frac{1}{3000s^{-1} + 2 \times 10^6 + 1s^{-2}}\right). \end{aligned} \tag{1.68}$$

As:

$$\lim_{s \to \infty}\left(\frac{1}{3000s^{-1} + 2 \times 10^6 + 1s^{-2}}\right) \neq 0, \tag{1.69}$$

then, the structure at infinity of the perturbation process without regulator is 2. As the structures at infinity of $H(s)$ and $H^d(s)$ are equal, using the Theorem 1.4, a regulator function $\nu = F\chi + Gd$ can be obtained such that:

$$C(sI - A - BF)^{-1}(BG + E) = 0, \tag{1.70}$$

where $F = \begin{bmatrix} f_1 & f_2 & f_3 \end{bmatrix}$ and G is a scalar. The parameters of the regulator function are as follows:

$$F = \begin{bmatrix} -0.5 & -0.5 & 1 \end{bmatrix}, \tag{1.71}$$

and:

$$G = 0. \tag{1.72}$$

So, the regulator function turns out to be:

$$\nu = \begin{bmatrix} -0.5 & -0.5 & 1 \end{bmatrix} \chi. \tag{1.73}$$

1.7.2 Perturbation Attenuation Via the Geometric Method Regulator

The invariant subspace $\mathcal{V}^*$ is calculated in this strategy. Start obtaining ker C:

$$\begin{bmatrix} 0 & 0 & 1 \end{bmatrix} \begin{bmatrix} \chi_1 \\ \chi_2 \\ \chi_3 \end{bmatrix} = 0. \tag{1.74}$$

Equation (1.74) implies that every vector in ker C should have coordinates satisfying:

$$\chi_3 = 0. \tag{1.75}$$

So,

$$\begin{bmatrix} \chi_1 & \chi_2 & \chi_3 \end{bmatrix}^T = p_1\chi_1 + p_2\chi_2, \tag{1.76}$$

where $p_1 = \begin{bmatrix} 1 & 0 & 0 \end{bmatrix}^T$ and $p_2 = \begin{bmatrix} 0 & 1 & 0 \end{bmatrix}^T$. So,

$$\ker\ C = \text{gen}\,\{p_1, p_2\} = \mathcal{V}^0. \tag{1.77}$$

As dim(ImB) = rank(B) = 1, then,

$$\text{Im}(B) = \text{gen}\,\{b_1\}, \tag{1.78}$$

where $b_1 = \begin{bmatrix} 1 & 0 & 1 \end{bmatrix}^T$. So,

$$\mathcal{S}_1 = \text{Im}\,B \oplus \mathcal{V}^0 = \text{gen}\,\{p_1, p_2, b_1\}. \tag{1.79}$$

To obtain $\chi \in A^{-1}(\mathcal{S}_1)$, consider:

$$A \begin{bmatrix} \chi_1 \\ \chi_2 \\ \chi_3 \end{bmatrix} = \begin{bmatrix} -0.0005\chi_1 \\ -0.0005\chi_2 \\ 0.0005\chi_1 + 0.0005\chi_2 - 0.001\chi_3 \end{bmatrix} \in \mathcal{S}_1, \tag{1.80}$$

Solving gives:

$$\begin{bmatrix} \chi_1 \\ \chi_2 \\ \chi_3 \end{bmatrix} = \begin{bmatrix} 0 \\ -2000 \\ -1000 \end{bmatrix}. \tag{1.81}$$

As $\dim\left(A^{-1}\mathcal{S}_1\right) = \dim(\ker A) \oplus \dim\left(\mathcal{S}_1 \cap \operatorname{Im} A\right) = 0 + 2 = 2$, then:

$$A^{-1}\mathcal{Z} = \text{gen}\left\{\begin{bmatrix} 0 \\ -2000 \\ 0 \end{bmatrix}, \begin{bmatrix} 0 \\ 0 \\ -1000 \end{bmatrix}\right\}. \tag{1.82}$$

So,

$$\begin{aligned} \mathcal{V}^1 &= \ker\ C \cap A^{-1}\left(\operatorname{Im} B + \mathcal{V}^0\right) \\ &= \ker\ C \cap A^{-1}\mathcal{S}_1 = \begin{bmatrix} 0 & 1 & 0 \end{bmatrix}^T. \end{aligned} \tag{1.83}$$

Now, making a similar procedure gives:

$$\begin{aligned} \mathcal{V}^2 &= \ker\ C \cap A^{-1}\left(\operatorname{Im} B + \mathcal{V}^1\right) \\ &= \ker\ C \cap A^{-1}\mathcal{S}_2 = \begin{bmatrix} 0 & 1 & 0 \end{bmatrix}^T. \end{aligned} \tag{1.84}$$

As $\mathcal{V}^1 = \mathcal{V}^2$, so,

$$\mathcal{V}^* = \text{gen}\left\{\begin{bmatrix} 0 \\ 1 \\ 0 \end{bmatrix}\right\}. \tag{1.85}$$

It can be seen that $\operatorname{Im} E \subset \mathcal{V}^* \oplus \operatorname{Im} B = \text{gen}\{\mathcal{V}^*, b_1\}$. As $\operatorname{Im} E \subset \mathcal{V}^* + \operatorname{Im} B$; then, there exist $F = \begin{bmatrix} f_1 & f_2 & f_3 \end{bmatrix}$ and G being a scalar such that the regulator function $\nu = F\chi + Gd$ attenuates the perturbations. Set:

$$(A + BF)\mathcal{V}^* = \mathcal{V}^*. \tag{1.86}$$

The constants f_{ij} are found to satisfy (1.86). F is obtained as:

$$F = \begin{bmatrix} -0.5 & -0.5 & 1 \end{bmatrix}, \tag{1.87}$$

so, the regulator function turns out to be:

$$\nu = \begin{bmatrix} -0.5 & -0.5 & 1 \end{bmatrix}\chi. \tag{1.88}$$

1.7.3 Simulation

For the simulations of the process, the initial conditions are $\chi_0 = \begin{bmatrix} 20 & 20 & 20 \end{bmatrix}^T$; meaning that the initial temperature of the tanks 1–3 is 20 °C. It is desired that the temperature of the water tank 3 remains at 20 °C. The perturbations are water with a temperature of $d_1 = 5$ °C, which flows in the tanks 1 and 2. As the water from these tanks flows into the tank 3, the temperatures of all three tanks should become cold if the perturbations are not attenuated. Figure 1.7 shows the regulator functions for the thermal process. The comparisons of the states of the thermal process are shown in Figure 1.8. Figure 1.9 shows the comparisons for the output of the thermal process.

From Figure 1.8, it can be seen that in spite of the temperature of the water of the tanks 1 and 2 is becoming cold, the temperature of the tank 3 remains at 20 °C when the regulator is used. From Figure 1.9, it can be observed that the output of the thermal process without regulator have perturbations, but the output of the thermal process with a regulator does not have perturbations, so the temperature of the tank 3 remains in 20 °C when the regulator is used.

Remark 1.6: From (1.73 and 1.88), it is shown that both strategies show that the perturbations are not required in the regulator function to be attenuated in the thermal process.

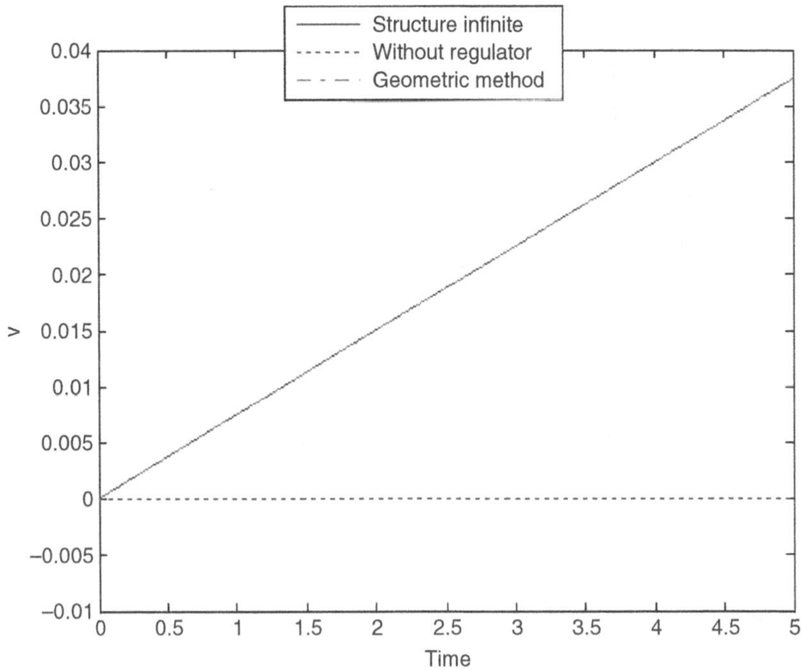

Figure 1.7 Regulator functions for the thermal process.

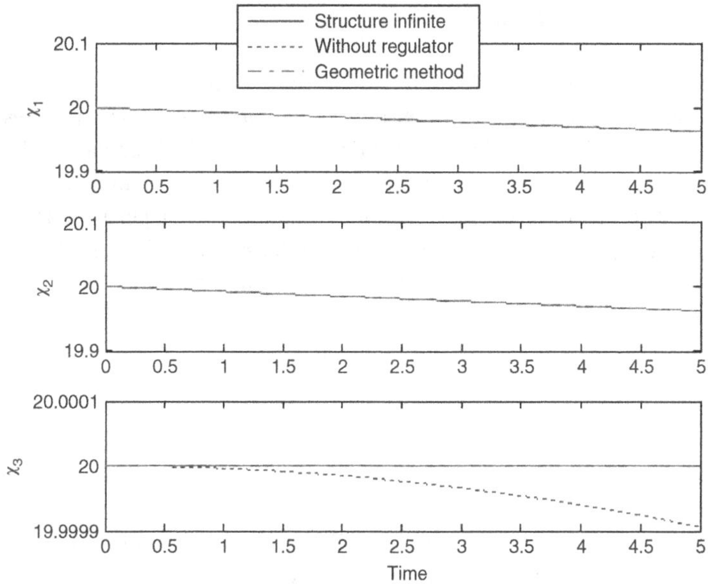

Figure 1.8 States of the thermal process.

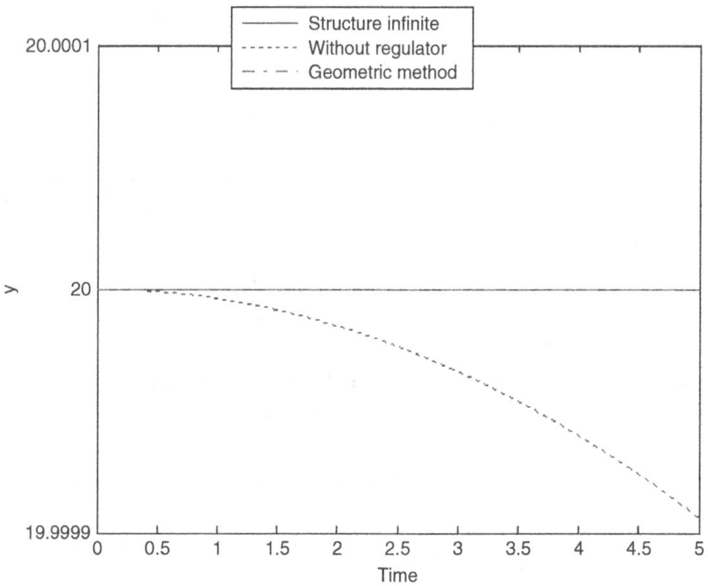

Figure 1.9 Output of the thermal process.

1.8 Concluding Remarks

The chapter has recommended two strategies for the perturbation attenuation, which act on the inputs and affect the outputs of the processes. The strategies were the structure at infinity regulator and the geometric method regulator. The two proposed strategies were used in a mechanical process and a thermal process, showing that the outputs of the processes without regulator have a perturbations, but the outputs of the processes with regulator do not have a perturbations. In the forthcoming work, the stability of the regulator will be warranted, and the perturbations will be attenuated for the processes, or the perturbations will be considered unknown and will be attenuated or decreased using the evolving intelligent processes.

References

1 Chen J., Sun F., Yin Y., Hu C., (2011), State feedback robust stabilisation for discrete-time fuzzy singularly perturbed systems with parameter uncertainty, IET Control Theory and Applications, 5 (10), 1195–1202.
2 Yang Z. J., Wang Y., Kanae S., (2011), New approach to an adaptive robust motion controller combined with a disturbance observer, IET Control Theory and Applications, 5 (10), 1203–1213.
3 Zhang R., Quan Q., Cai K. Y., (2011), Attitude control of a quadrotor aircraft subject to a class of time-varying disturbances, IET Control Theory and Applications, 5 (9), 1140–1146.
4 Dunnigan M. W., Wronka C. M., (2011), Comparison of control techniques for a robotic manipulator with base disturbances, IET Control Theory and Applications, 5 (8), 999–1012.
5 Wu H. N., Li H. X., (2011), Robust adaptive L_∞-gain neural filtering for non-linear systems in the presence of bounded disturbances, IET Control Theory and Applications, 5 (4), 630–639.
6 Li H., Sun Z., Liu H., Sun F., Deng J., (2011), State feedback integral control of networked control systems with external disturbance, IET Control Theory and Applications, 5 (2), 283–290.
7 Wonham W. M., (1985), Linear Multivariable Control: A Geometric Approach. Springer-Verlag.
8 Zwart H. J., (1989), Geometric Theory for Infinite Dimensional Systems. (Lecture Notes in Control and Information Sciences 115.) Springer-Verlag.
9 Basile G., Marro G., (1991), Controlled and Conditioned Invariants in Linear System Theory. ISBN:0-13172974-8. Prentice Hall.

10 Rabah R., Malabre M., (1999), On the structure at infinity of linear delay systems with application to the disturbance decoupling problem, Kybernetika, 35 (6), 668–680.

11 Rabah R., Malabre M., (1993), Structure at infinity, model matching and disturbance rejection for linear systems with delays, Kybernetika, 29.

12 Malabre M., Kucera V., (1984), Infinite structure and exact model matching problem: a geometric approach. IEEE Transactions on Automatic Control, 3, 266–268.

13 Rubio J. J., Figueroa M., Perez-Cruz J. H., Bejarano F. J., (2012), Geometric approach and structure at infinite controls for the disturbance rejection, IET Control Theory and Applications, 6 (16), 2528–2537.

14 Commault C., Dion J. M., (1992), Comprehensive introduction to the geometric theory of linear multivariable systems, IEEE Transactions on Education, 35 (1), 92–97.

15 Dion J. M., Commault C., Trinh D. H., (2008), Sensor classification for the disturbance rejection by measurement feedback problem, Proceedings of the 17th World Congress The International Federation of Automatic Control Seoul, Korea, 11299–11303.

16 Eves H. W., (1968), Funciones de Variable Compleja, Tomo II, C.E.C.S.A.

17 Chen C. T., (1999), Linear Systems Theory and Design, third edition, ISBN 0-19-511777-8.

2

Structure Regulator for the Perturbation Attenuation in Two Processes

2.1 Introduction

The perturbations are undesired signals presented as inputs of a process, which influence the outputs. This problem has been studied in many industrial applications. The perturbation attenuation is desired because, in the contrary case, the regulators or plants could be damaged, causing accidents, or unnecessary costs. So, a regulator to attenuate perturbations is an important and current issue in the industrial applications.

There are some interesting works about regulators. The problems of regulator and filtering of observer form for linear and nonlinear models with perturbations are investigated in [1–3]. A theoretical approach of an optimal regulator is recommended in [4]. In [5], the stabilization for the Schrödinger equation with perturbations is considered. The optimal tracking of linear plants is investigated in [6]. In [7], the H-infinity feedback regulators of disturbed processes are recommended. In [8], robust observers of disturbed processes are designed. The trajectory tracking of dead-zone nonlinear processes is focused in [9]. In [10, 11], the trajectory tracking problem of direct current motors is considered. The robust output feedback regulator of disturbed processes is investigated in [12]. In [13], the position tracking for a tank regulator process with perturbations is investigated. Most of the processes require the perturbation attenuation; so, a regulator could overcome this problem.

There exists an interesting strategy for the perturbation attenuation: the structure at infinity regulator; it is a condition where the transfer functions of the process with the inputs or perturbations must be equal for the existence of the perturbation attenuation. There is some investigation about this issue. Some mathematical aspects of the structure at infinity are studied in [14]. In [15, 16], square invertible linear processes are decoupled by static state feedbacks. The structure at infinity of plants with input and output spaces is discussed in [17]. In [18], the structure at infinity for linear discrete time periodic processes is proposed. The solvability of nonlinear decoupling problems through the notion of structure at infinity is studied in [19]. In [20, 21], the structure at infinity strategy is explained. The structure at infinity strategy is employed for the perturbation attenuation in [8, 22]. In [23], the structure at infinity of the closed loop state space processes is examined. The structure at infinity of a structured process is developed in [24]. In [25], relations between some strict equivalence invariants and pole-zero structure at infinity are established. The structure at infinity regulator requires to satisfy two conditions for the perturbation attenuation: (1) the perturbations must be known to be employed in the regulator law and (2) the input transfer function must be equal with the perturbations transfer function. Since it is hard to satisfy the two conditions, a new regulator should be designed.

In this chapter, the structure regulator is recommended for the perturbation attenuation of two processes. The structure regulator improves the structure at infinity regulator as follows: (1) the knowledge of the perturbations is not required in the recommended strategy because it is not utilized in the regulator law and (2) the second condition required by the structure at infinity regulator is not employed in the structure regulator because the stability and convergence of the errors are warranted.

Finally, the recommended strategy is compared with the structure at infinity strategy for the perturbation attenuation in a three-phase circuit and a plotter. The three-phase circuit is employed in the electric power distribution and the plotter is utilized for printing vector graphics.

The remainder of the chapter is organized as follows. The structure at infinity regulator for the perturbation attenuation is explained in Section 2.2. In Section 2.3, the structure regulator for the perturbation attenuation, its stability, and its convergence analysis are recommended. The recommended strategy is compared with the structure at infinity strategy to attenuate the perturbations for a three-phase circuit in Section 2.4, and for a plotter in Section 2.5. In Section 2.6, the conclusion and future research are described.

2.2 Structure at Infinity Strategy for the Perturbation Attenuation

In this section, the structure at infinity strategy for the perturbation attenuation is described.

The perturbations process is:

$$\begin{aligned} \dot{\chi}_i &= A_i\chi_i + B_i\nu_i + E_i d_i, \\ y_i &= C_i\chi_i, \end{aligned} \tag{2.1}$$

where $\chi_i \in \mathcal{Z}_i \subseteq \Re^n$ are the states, $\nu_i \in \mathcal{V}_i \subseteq \Re^m$ are the inputs, $y_i \in \mathcal{Y}_i \subseteq \Re^p$ are the outputs, and $d_i \in \mathcal{D}_i \subseteq \Re^l$ are the perturbations. A_i, B_i, C_i, and E_i are the matrices which represent the transformations $A_i : \mathcal{Z}_i \to \mathcal{Z}_i$, $B_i : \mathcal{V}_i \to \mathcal{Z}_i$, $C_i : \mathcal{Z}_i \to \mathcal{Y}_i$, $D_i : \mathcal{V}_i \to \mathcal{Y}_i$, and $E_i : \mathcal{D}_i \to \mathcal{Z}_i$, respectively.

If the structure at infinity of the regulator process and perturbations process are equal, then the perturbations can be attenuated. So, there exists a regulator law for the perturbation attenuation.

The transfer function of the regulator process is:

$$T_\nu(s) = C_i(sI - A_i)^{-1}B_i. \tag{2.2}$$

The transfer function of the perturbations process is:

$$T_d(s) = C_i(sI - A_i)^{-1}E_i. \tag{2.3}$$

Theorem 2.1 describes the structure at infinity strategy.

Theorem 2.1: *If the structure at infinity in $T_\nu(s)$ of (2.2) and in $T_d(s)$ of (2.3) are equal, then the perturbations can be attenuated in the processes (2.1) by employing the following regulator law [8, 20–22]:*

$$\nu_i = F_i\chi_i + G_i d_i, \tag{2.4}$$

where

$$C_i(sI - A_i - B_iF_i)^{-1}(B_iG_i + E_i) = 0. \tag{2.5}$$

2.3 Structure Regulator for the Perturbation Attenuation

In this section, the structure regulator design, its stability, and its convergence are recommended for the perturbation attenuation as the main benefits of this chapter.

The perturbations process is as the follows:

$$\begin{aligned} \dot{\chi} &= A_c\chi + B_c\nu + E_c d, \\ y &= C_c\chi + D_c\nu, \end{aligned} \tag{2.6}$$

where $\chi \in \mathcal{Z} \subseteq \Re^n$ are the states, $\nu \in \mathcal{V} \subseteq \Re^m$ are the inputs, $y \in \mathcal{Y} \subseteq \Re^p$ are the outputs, and $d \in \mathcal{D} \subseteq \Re^l$ are the perturbations. A_c, B_c, C_c, D_c, and E_c are the matrices which represent the transformations $A_c : \mathcal{Z} \to \mathcal{Z}$, $B_c : \mathcal{V} \to \mathcal{Z}$, $C_c : \mathcal{Z} \to \mathcal{Y}$, $D_c : \mathcal{V} \to \mathcal{Y}$, and $E_c : \mathcal{D} \to \mathcal{Z}$, respectively.

2.3.1 Regulator Design

The proposed regulator can be seen in Figure 2.1. Note that the perturbations are unknown in the regulator.

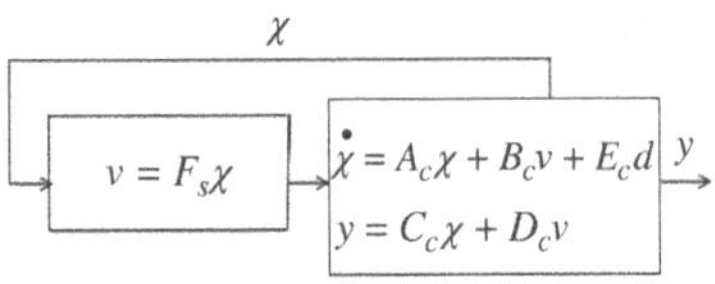

Figure 2.1 Regulator with unknown perturbations.

Theorem 2.2 presents the first main benefit of this study, the perturbation attenuation regulator.

Theorem 2.2: *The perturbations can be attenuated in the processes (2.6) by employing the following structure regulator law:*

$$\nu = F_s\chi, \tag{2.7}$$

such that:

$$T_s(s) = C_s(sI - A_s)^{-1}B_s = 0, \tag{2.8}$$

where $F_s : \mathcal{Z} \to \mathcal{V}$, $A_s = A_c + B_cF_s$, $B_s = E_c$, *and* $C_s = C_c + D_cF_s$. $T_s(s)$ *is the transfer function of the outputs and perturbations.*

Proof. See [26] for the proof. ■

In this part of the study, an analysis of the modifications of (2.2) and (2.3) in the structure at infinity approach of Theorem 2.1 to be applied to the processes (2.6) is mentioned with a corollary.

Corollary 2.1: *The condition which says that the structure at infinity in* $T_\nu(s)$ *of (2.9) and in* $T_d(s)$ *of (2.10) must be equal, is not required in the structure regulator for the perturbation attenuation, where the transfer function of the regulator process is obtained when* $d = 0$:

$$T_\nu(s) = \frac{y(s)}{\nu(s)} = C_c(sI - A_c)^{-1}B_c + D_c, \tag{2.9}$$

and the transfer function of the perturbations process is obtained when $\nu = 0$:

$$T_d(s) = \frac{y(s)}{d(s)} = C_c(sI - A_c)^{-1}E_c. \tag{2.10}$$

Remark 2.1: It is important to note that the processes of this chapter are more difficult to regulator than the previous proccsses because the processes of (2.6) have the additional term $D_c\nu$, which is not considered in the processes of (2.1).

2.3.2 Stability Analysis

In this section, states of the structure regulator for the perturbation attenuation are warranted to be uniform stable based on the solution of the Lyapunov strategy. It is the second main benefit of this chapter.

The states equation can be written as follows:

$$\dot{\chi} = A_s\chi + B_s d, \tag{2.11}$$

where $A_s = A_c + B_c F_s$ and $B_s = E_c$.

Theorem 2.3 shows the stability of the proposed regulator.

Theorem 2.3: *The state χ of the structure regulator (2.7) used for the perturbation attenuation in the processes (2.6) is uniform stable; so, the states χ satisfy:*

$$\|\chi\|^2 \leq \gamma e^{-\beta t}\|\chi_0\|^2 + \xi, \tag{2.12}$$

where χ_0 is the initial condition of χ, $\gamma = \frac{\lambda_{\max}(O)}{\lambda_{\min}(O)}$, $\xi = \frac{\alpha}{\beta\lambda_{\min}(O)}$, $\beta = \lambda_{\min}(UO^{-1})$, $\alpha = \frac{1}{\theta}B_s^T O B_s \overline{d}^2$, $|d^2| \leq \overline{d}^2$, $0 < \gamma \in \mathfrak{R}$, $0 < \beta \in \mathfrak{R}$, $0 < \xi \in \mathfrak{R}$, $\|\cdot\|$ is the norm and $|\cdot|$ is the absolute value, $O \in \mathfrak{R}^{n\times n}$ and $U \in \mathfrak{R}^{n\times n}$ are positive matrices, which satisfy the following equation:

$$A_s^T O + OA_s + \theta O = -U, \tag{2.13}$$

where A_s is given in (2.11), $0 < \theta \in \mathfrak{R}$ is selected by the designer.

Proof. See [26] for the proof. ■

2.3.3 Convergence Analysis

The following theorem presents the convergence of the states in the structure regulator for the perturbation attenuation. It is the third main benefit of this work.

Theorem 2.4: *If the real parts of the eigenvalues in A_s are negative, then the states of the regulator (2.7) used for the perturbation attenuation of the process (2.6) will converge to:*

$$\lim_{t\to\infty} |\chi| \leq \overline{\chi} \tag{2.14}$$

where $\overline{\chi} = \left|-A_s^{-1}B_s\overline{d}\right|$, $|d| \leq \overline{d}$, A_s^{-1} is the pseudo-inverse of A_s, A_s and B_s are given in (2.11), d is given in (2.6), $|\cdot|$ is the absolute value.

Proof. See [26] for the proof. ■

Remark 2.2: The structure regulator (2.7) requires to satisfy the condition (2.8) of Theorem 2.2 to guarantee the perturbation attenuation, the condition (2.13) of Theorem 2.3 to assure the stability of its error, and the condition (2.14) of Theorem 2.4 to guarantee the convergence of its error. So, the three conditions are required by the structure regulator to obtain an acceptable perturbation attenuation and regulation.

Remark 2.3: There are three differences between the structure regulator of this chapter and the structure at infinity regulator of [8, 17, 20–23]. The first difference is that the perturbations are not utilized in the regulator law of the structure regulator, while the perturbations are required in the regulator law for the structure at infinity regulator. The second difference is that the stability of the structure regulator is warranted, while the stability of the structure at infinity is not warranted. And the third difference is that the convergence for the structure regulator is warranted, while the convergence for the structure at infinity is not warranted.

Remark 2.4: Only the upper bounds of the perturbations are utilized in the stability and convergence analysis, but they are not utilized in the regulator laws.

Remark 2.5: The gains of the structure regulator are selected to satisfy three benefits; the perturbation attenuation of the Theorem 2.2, the stability of the Theorem 2.3, and the convergence of the errors of the Theorem 2.4.

In the following two sections, the root mean square error (RMSE) will be used for the comparisons, it is given as follows:

$$J_y = \left(\frac{1}{T} \int_0^T y^2 d\tau \right)^{\frac{1}{2}}, \tag{2.15}$$

where $y^2 = y^2$ or $y^2 = y_1^2 + y_2^2$ for one or two outputs.

2.4 Perturbation Attenuation in a Three-phase Electric Circuit

Figure 2.2 shows the three-phase circuit. The three-phase circuit is employed in the electric power distribution. It has three sources, three resistances, and three coils [27].

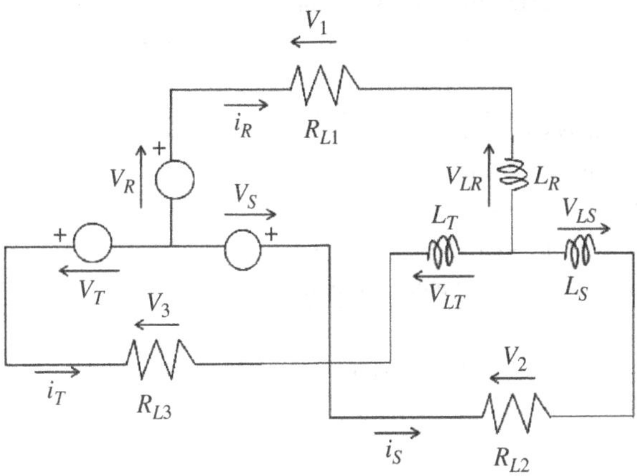

Figure 2.2 The three-phase electric circuit.

The state variables are $\chi_1 = i_R$, $\chi_2 = i_S$, the inputs are $\nu_1 = V_R$, $\nu_2 = V_S$, $\nu_3 = V_T$ and the outputs are $y_1 = -V_1 + V_R = -R_{L1}\chi_1 + \nu_1$, $y_2 = V_2 + V_S = R_{L2}\chi_2 + \nu_2$. The analytic model is given as follows:

$$\begin{aligned}
\dot{\chi}_1 &= -\frac{(L_S + L_T)\,R_{L1} + L_S R_{L3}}{L_{aux}}\chi_1 - \frac{L_T R_{L2} - L_S R_{L3}}{L_{aux}}\chi_2 \\
&\quad + \frac{L_S + L_T}{L_{aux}}\nu_1 - \frac{L_T}{L_{aux}}\nu_2 - \frac{L_S}{L_{aux}}\nu_3, \\
\dot{\chi}_2 &= -\frac{L_T R_{L1} - L_R R_{L3}}{L_{aux}}\chi_1 - \frac{(L_R + L_T)\,R_{L2} + L_R R_{L3}}{L_{aux}}\chi_2 \\
&\quad - \frac{L_T}{L_{aux}}\nu_1 + \frac{L_R + L_T}{L_{aux}}\nu_2 - \frac{L_R}{L_{aux}}\nu_3, \\
y_1 &= -R_{L1}\chi_1 + \nu_1, \\
y_2 &= R_{L2}\chi_2 + \nu_2,
\end{aligned} \tag{2.16}$$

where $L_{aux} = L_R L_S + L_R L_T + L_S L_T$. Table 2.1 shows the parameter descriptions for the three-phase circuit. Table 2.2 shows the parameter values for the three-phase circuit. The perturbations process of the three-phase circuit is represented as (2.6), where $d = d$ are the perturbations, $\nu = [\nu_1, \nu_2, \nu_3]^T$ are the regulator inputs, $\chi = [\chi_1, \chi_2]^T$ are the states, and $y = [y_1, y_2]^T$ are the outputs.

Table 2.1 Parameter descriptions for the three-phase circuit.

Parameter	Description
L_R, L_S, L_T	Inductances 1, 2, and 3 in H
R_{L1}, R_{L2}, R_{L3}	Resistances 1, 2, and 3 in Ω
V_1, V_2, V_3	Voltage of the resistances 1, 2, and 3 in V
V_R, V_S, V_T	Voltage of the sources 1, 2, and 3 in V
V_{LR}, V_{LS}, V_{LT}	Voltage of the inductances 1, 2, and 3 in V

Table 2.2 Parameter values for the three-phase circuit.

Parameter	Value	Parameter	Value
L_R	10×10^{-3} H	R_{L1}	4.7 Ω
L_S	10×10^{-3} H	R_{L2}	5.6 Ω
L_T	10×10^{-3} H	R_{L3}	6.8 Ω

Using parameter values of Table 2.2, the matrices of the perturbations process (2.6) are follows:

$$
\begin{aligned}
A_c &= \begin{bmatrix} -540 & -40 \\ -70 & -600 \end{bmatrix}, \\
B_c &= \begin{bmatrix} \frac{200}{3} & -\frac{100}{3} & -\frac{100}{3} \\ -\frac{100}{3} & \frac{200}{3} & -\frac{100}{3} \end{bmatrix}, \\
C_c &= \begin{bmatrix} -4.7 & 0 \\ 0 & 5.6 \end{bmatrix}, \\
D_c &= \begin{bmatrix} 1 & 0 & 0 \\ 0 & 1 & 0 \end{bmatrix}, \\
E_c &= \begin{bmatrix} 1 & 1 \end{bmatrix}^T.
\end{aligned} \tag{2.17}
$$

2.4.1 Structure at Infinity Regulator Design

From the Corollary 2.1, the transfer function $T_\nu(s)$ of (2.9) is given as follows:

$$T_\nu(s) = \begin{bmatrix} \frac{-6266.7 - 313.33(s+60)}{396s + s^2 + 25520} + 1 & \frac{12533 + 156.67(s+60)}{396s + s^2 + 25520} \\ \frac{50027 - 186.67(s+336)}{396s + s^2 + 25520} & \frac{-25013 + 373.33(s+336)}{396s + s^2 + 25520} + 1 \\ \frac{-6266.7 + 156.67(s+60)}{396s + s^2 + 25520} & \\ \frac{-25013 - 186.67(s+336)}{396s + s^2 + 25520} & \end{bmatrix}. \tag{2.18}$$

The transfer function $T_d(s)$ of (2.10) is given as follows:

$$T_d(s) = \begin{bmatrix} \frac{188.0 - 4.7(s+60)}{396s + s^2 + 25520} \\ \frac{750.4 + 5.6(s+336)}{396s + s^2 + 25520} \end{bmatrix}. \tag{2.19}$$

Since the structure at infinity of $T_\nu(s)$ of (2.18) and $T_d(s)$ of (2.19) is not equal, F_i can not be found using this strategy. So, the structure at infinity regulator law is as follows:

$$\nu_i = \begin{bmatrix} 0 & 0 \\ 0 & 0 \\ 0 & 0 \end{bmatrix} \chi_i. \tag{2.20}$$

2.4.2 Structure Regulator Design

From (2.7) of the Theorem 2.2, a structure regulator law $\nu = F_s\chi$ can be obtained such that using (2.8):

$$T_s(s) = C_s(sI - A_s)^{-1}B_s + D_s = 0, \tag{2.21}$$

where $F_s = \begin{bmatrix} f_1 & f_4 \\ f_2 & f_5 \\ f_3 & f_6 \end{bmatrix}$. Substituting F_s in (2.21) given as the following:

$$\left\{\left(\begin{bmatrix}-4.7 & 0\\ 0 & 5.6\end{bmatrix}+\left(\begin{bmatrix}1 & 0 & 0\\ 0 & 1 & 0\end{bmatrix}\begin{bmatrix}f_1 & f_4\\ f_2 & f_5\\ f_3 & f_6\end{bmatrix}\right)\right)\right.$$
$$*\left(sI-\begin{bmatrix}-540 & -40\\ -70 & -600\end{bmatrix}-\left(\begin{bmatrix}\frac{200}{3} & -\frac{100}{3} & -\frac{100}{3}\\ -\frac{100}{3} & \frac{200}{3} & -\frac{100}{3}\end{bmatrix}\begin{bmatrix}f_1 & f_4\\ f_2 & f_5\\ f_3 & f_6\end{bmatrix}\right)\right)^{-1} \tag{2.22}$$
$$\left.*\left(\begin{bmatrix}1\\ 1\end{bmatrix}\right)\right\}=\begin{bmatrix}0.0\\ 0.0\end{bmatrix},$$

Since this system is the multiplication of three terms, its solution is to find F_s to obtain zero in one of the terms, the first term is selected as follows:

$$\left(\begin{bmatrix}-4.7 & 0\\ 0 & 5.6\end{bmatrix}+\left(\begin{bmatrix}1 & 0 & 0\\ 0 & 1 & 0\end{bmatrix}\begin{bmatrix}f_1 & f_4\\ f_2 & f_5\\ f_3 & f_6\end{bmatrix}\right)\right)=\begin{bmatrix}0 & 0\\ 0 & 0\end{bmatrix}. \tag{2.23}$$

Making operations given as follows:

$$\begin{bmatrix}f_1-4.7 & f_4\\ f_2 & f_5+5.6\end{bmatrix}=\begin{bmatrix}0 & 0\\ 0 & 0\end{bmatrix}, \tag{2.24}$$

where the solution is given as follows:

$$\begin{aligned}f_1&=4.7,\\ f_2&=0.0,\\ f_4&=0.0,\\ f_5&=-5.6,\\ f_3&=0.0,\\ f_6&=0.0.\end{aligned} \tag{2.25}$$

So, F_s is found as follows:

$$F_s=\begin{bmatrix}4.7 & 0\\ 0 & -5.6\\ 0 & 0\end{bmatrix}. \tag{2.26}$$

So, the structure regulator law is as follows:

$$\nu = \begin{bmatrix} 4.7 & 0 \\ 0 & -5.6 \\ 0 & 0 \end{bmatrix} \chi. \tag{2.27}$$

2.4.3 Comparisons

In this section, the structure regulator of this chapter named structure is compared with the structure at infinity strategy of [8, 22] named structure at infinity. The mathematical model of the three-phase circuit is given by (2.16). The benefit of the regulators is to attenuate the perturbations in the outputs of the three-phase circuit. The initial conditions are $\chi_0 = [1 \times 10^{-3}, 1 \times 10^{-3}]^T$; it is desired that the three-phase circuit does not have output voltages if it receives input voltages. The perturbations are the incoming voltage, which is defined as follows $d = 0.25\sin(20t)$ V from 0 s to 4 s, t is the time in s.

The structure at infinity regulator of [8, 22] with the regulator law (2.20) is employed for the perturbation attenuation in the three-phase circuit (2.6) and (2.17).

The structure regulator of this study with the regulator law (2.27) of Theorem 2.2 is utilized for the perturbation attenuation in the three-phase circuit (2.6) and (2.17). The condition of Theorem 2.3 is satisfied with $O = \begin{bmatrix} 2.0599 \times 10^{-3} & 1.5052 \times 10^{-4} \\ 1.5052 \times 10^{-4} & 5.3665 \times 10^{-4} \end{bmatrix}$, $U = \begin{bmatrix} 1 & -9.5975 \times 10^{-6} \\ -9.5975 \times 10^{-6} & 0.99998 \end{bmatrix}$; so, the other values are $\overline{d}^2 = 0.25^2 = 0.0625$, $\alpha = \frac{1}{\theta} B_s^T O B_s \overline{d}^2 \frac{1}{1 \times 10^{-0}} (2.8976 \times 10^{-3})(0.0625) = 1.811 \times 10^{-4}$, $\gamma = \frac{\lambda_{\max}(O)}{\lambda_{\min}(O)} = \frac{2.0746 \times 10^{-3}}{5.2192 \times 10^{-4}} = 3.9749$, $\beta = \lambda_{\min}(UO^{-1}) = 482.01$, and $\xi = \frac{\alpha}{\beta \lambda_{\min}(O)} = \frac{1.811 \times 10^{-4}}{482.01(5.2192 \times 10^{-4})} = 7.1988 \times 10^{-4}$. The condition the Theorem 2.4 is satisfied with the eigenvalues of A_s are -274.22, -925.78, where all the real parts are negative; so, the second state of the regulator in the process will converge to $\lim_{t\to\infty} |\chi_1| \leq 1.1029 \times 10^{-3}$, and $\lim_{t\to\infty} |\chi_2| \leq 0$.

Figures 2.3–2.6 show input and perturbations, states, output, and RMSE of the structure and structure at infinity applied for the perturbation attenuation in the three-phase circuit, respectively. Table 2.3 shows the RMSE of (2.15).

From Figures 2.3 and 2.4, it can be seen that in structure the regulator law is varying with the time; this fact is used for the perturbation attenuation and influences the movement of the states, while in structure at infinity the regulator law does not have a behavior. From Table 2.3 and Figures 2.5 and 2.6, it can be observed that structure achieves better accuracy when compared with structure at infinity

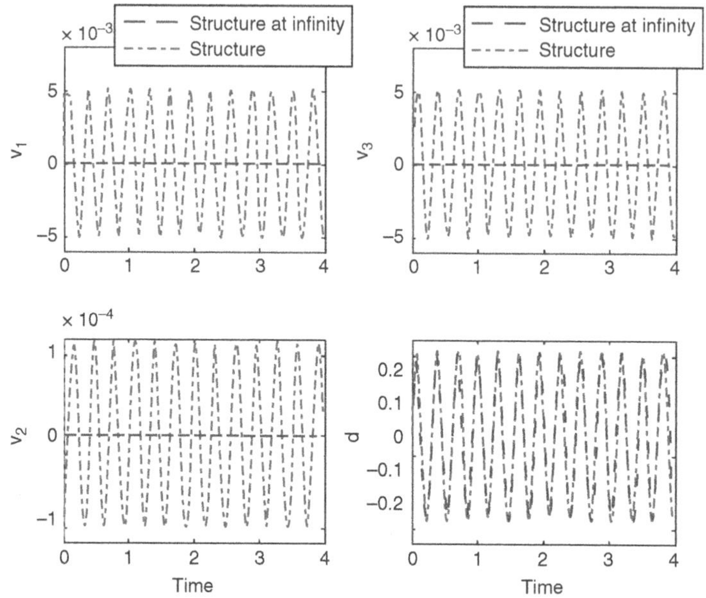

Figure 2.3 Inputs and perturbations of the three-phase circuit.

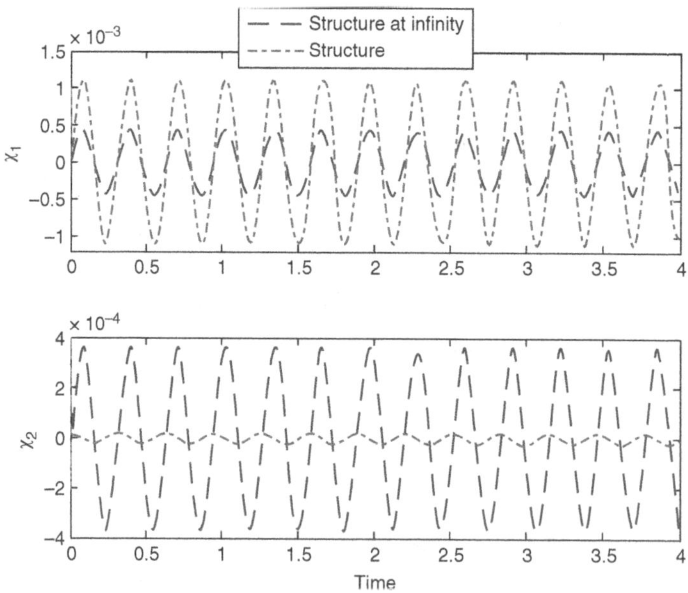

Figure 2.4 States of the three-phase circuit.

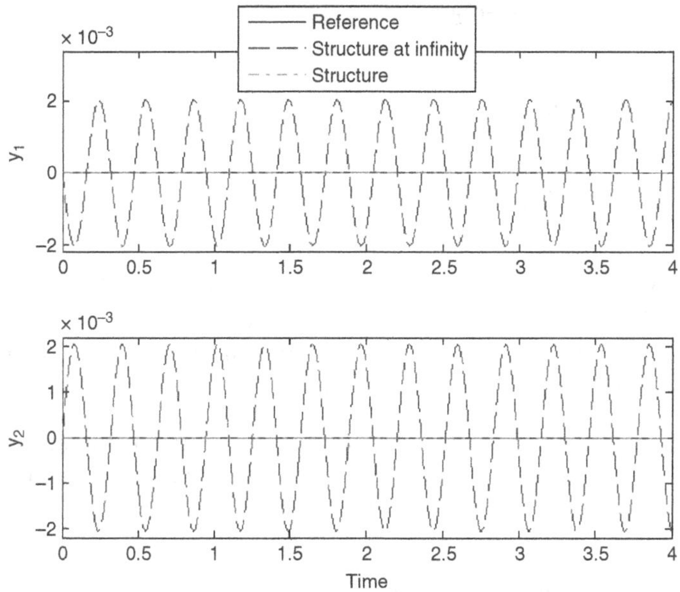

Figure 2.5 Outputs of the three-phase circuit.

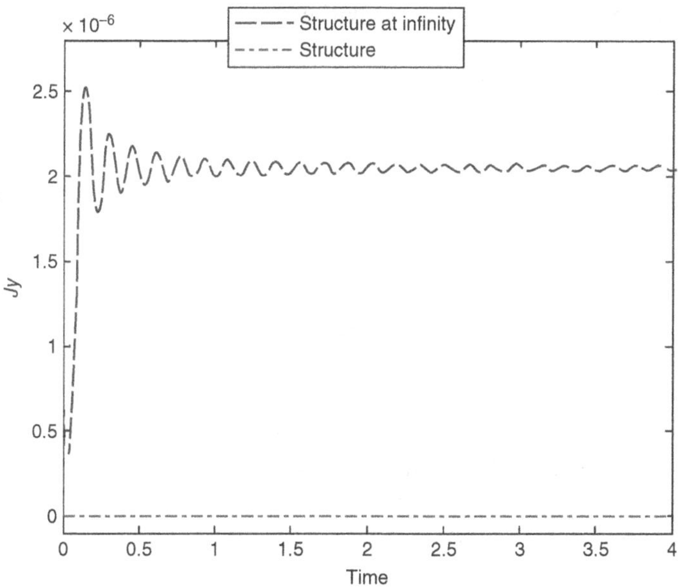

Figure 2.6 RMSE of the three-phase circuit.

Table 2.3 Comparisons for the three-phase circuit.

Strategies	Jy
Structure at infinity	2.0947×10^{-6}
Structure	0

for the perturbation attenuation, because in the first the outputs follow better the reference outputs than in the second, and because the RMSE is smaller for the first than for the second. However, structure obtains its benefit, the perturbation attenuation in the outputs of the three-phase circuit.

Remark 2.6: From (2.27), the proposed strategy shows that the perturbations are not employed in the structure regulator law to be attenuated in the three-phase circuit.

2.5 Perturbation Attenuation in a Plotter

Figure 2.7 shows the plotter. The plotter is a computer printer for printing vector graphics. It has a CD motor, a disc 1, a disc 2, a flywheel, a disc 3, and a poppet [27].

The states are $\chi_1 = i$, $\chi_2 = \theta_1$, $\chi_3 = \omega_1$, the input is $\nu = V_e$, and the output is $y = R_m i + V_e = R_m \chi_1 + \nu$, the mathematical model is as follows:

$$
\begin{aligned}
\dot{\chi}_1 &= -\frac{R_m}{L_m}\chi_1 - \frac{K_\nu}{L_m}\chi_3 + \frac{1}{L_m}\nu, \\
\dot{\chi}_2 &= \chi_3, \\
\dot{\chi}_3 &= \frac{K_i}{J_1 + \frac{r_1^2}{r_2^2}(J_2 + M)}\chi_1 - \frac{K_1 + \frac{r_1^2}{r_2^2}K_2}{J_1 + \frac{r_1^2}{r_2^2}(J_2 + M)}\chi_2 - \frac{B_1 + \frac{r_1^2}{r_2^2}B_2}{J_1 + \frac{r_1^2}{r_2^2}(J_2 + M)}\chi_3, \\
y &= R_m \chi_1 + \nu.
\end{aligned}
\tag{2.28}
$$

Table 2.4 shows the parameter descriptions for the plotter. Table 2.5 shows the parameter values for the plotter. The perturbations process of the plotter is represented as (2.6), where d are the perturbations, $\nu = \upsilon$ is the regulator input, $\chi = [\chi_1, \chi_2, \chi_3]^T$ are the states, and $y = y$ is the output.

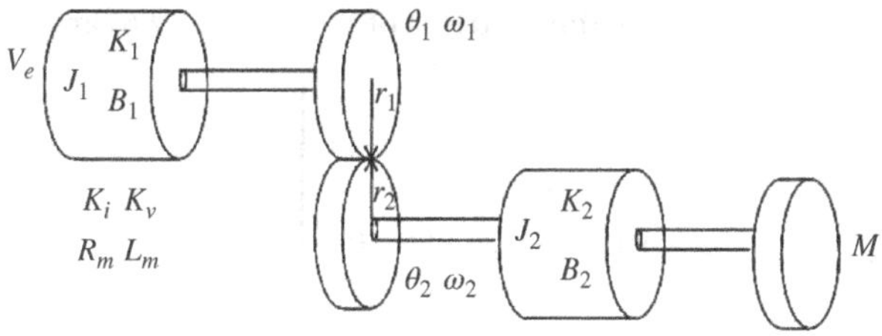

Figure 2.7 The plotter.

Table 2.4 Parameters descriptions for the plotter.

Parameter	Description
R_m, L_m	Resistance and inductance of the motor
J_1, J_2	Inertia of the discs 1 and 2 in kg m^2
B_1, B_2	Viscous friction of the discs 1 and 2 in kg m^2 rad/s
M	Mass of the pen nib in kg
V_e, i	Input voltage and current in the motor
θ_1, θ_2	Angle of the discs 1 and 2 in rad
ω_1, ω_2	Velocity of the discs 1 and 2 in rad/s
K_i	Motor magnetic flux of the motor in Wb
K_v	Back emf of the motor in Vs/rad
K_1, K_2	Spring effect of the discs 1 and 2 in kg m^2/s^2
r_1, r_2	Radius of the discs 1 and 2 in m

Table 2.5 Parameter values for the plotter.

Parameter	Value	Parameter	Value
R_m	1 Ω	L_m	10×10^{-3} H
J_1	0.01213 kg m^2	J_2	0.01213 kg m^2
B_1	0.1 kg m^2 rad/s	B_2	0.1 kg m^2 rad/s
K_i	0.01 Wb	K_v	0.45 V s/rad
K_1	1×10^{-3} kg m^2/s^2	K_2	1×10^{-3} kg m^2/s^2
r_1	0.1 m	r_2	0.1 m
M	0.5 kg		

Using the parameter values of Table 2.5, the matrices of the process (2.6) are:

$$A_c = \begin{bmatrix} -100 & 0 & -45.0 \\ 0 & 0 & 1 \\ 4.9372\times 10^{-3} & -9.8743\times 10^{-4} & -9.8743\times 10^{-2} \end{bmatrix},$$
$$B_c = \begin{bmatrix} 100 & 0 & 0 \end{bmatrix}^T,$$
$$C_c = \begin{bmatrix} 1 & 0 & 0 \end{bmatrix},$$
$$D_c = 1,$$
$$E_c = \begin{bmatrix} 1 & 0 & 0 \end{bmatrix}^T. \tag{2.29}$$

2.5.1 Structure at Infinity Regulator Design

From the corollary 1, the transfer function $T_\nu(s)$ of (2.9) is obtained as follows:

$$T_\nu(s) = 100\frac{9.8743\times 10^{6}s + 1.0\times 10^{8}s^2 + 98743}{1.0097\times 10^{9}s + 1.0010\times 10^{10}s^2 + 1.0\times 10^{8}s^3 + 9.8743\times 10^{6}} + 1. \tag{2.30}$$

The transfer function $T_d(s)$ of (2.10) is obtained as follows:

$$T_d(s) = \frac{9.8743\times 10^{6}s + 1.0\times 10^{8}s^2 + 98743}{1.0097\times 10^{9}s + 1.0010\times 10^{10}s^2 + 1.0\times 10^{8}s^3 + 9.8743\times 10^{6}}. \tag{2.31}$$

Since the structure at infinity of $T_\nu(s)$ of (2.30) and $T_d(s)$ of (2.31) is not equal, F_i cannot be found using this strategy. So, the structure at infinity regulator law is as follows:

$$\nu_i = \begin{bmatrix} 0 & 0 & 0 \end{bmatrix} \chi_i. \tag{2.32}$$

2.5.2 Structure Regulator Design

From (2.7) of the Theorem 2.2, a regulator function $\nu = F_s\chi$ can be obtained such that using (2.8):

$$T_s(s) = C_s(sI - A_s)^{-1}B_s + D_s = 0, \tag{2.33}$$

where $F_s = \begin{bmatrix} f_1 & f_2 & f_3 \end{bmatrix}$. Substituting F_s in (2.33) given as follows:

$$\left\{\left(\begin{bmatrix}1 & 0 & 0\end{bmatrix}+\left(1\begin{bmatrix}f_1 & f_2 & f_3\end{bmatrix}\right)\right)\right.$$

$$*\left(sI-\begin{bmatrix}-100 & 0 & -45.0\\ 0 & 0 & 1\\ 4.9372\times10^{-3} & -9.8743\times10^{-4} & -9.8743\times10^{-2}\end{bmatrix}-\left(\begin{bmatrix}100\\0\\0\end{bmatrix}\begin{bmatrix}f_1 & f_2 & f_3\end{bmatrix}\right)\right)^{-1}$$

$$\left.*\left(\begin{bmatrix}1\\0\\0\end{bmatrix}\right)\right\}=0, \tag{2.34}$$

where $I=\begin{bmatrix}1 & 0 & 0\\0 & 1 & 0\\0 & 0 & 1\end{bmatrix}$. Since this system is the multiplication of three terms, its solution is to find F_s to obtain zero in one of the terms. The first term is selected as follows:

$$\left(\begin{bmatrix}1 & 0 & 0\end{bmatrix}+\left(1\begin{bmatrix}f_1 & f_2 & f_3\end{bmatrix}\right)\right)=\begin{bmatrix}0 & 0 & 0\end{bmatrix}. \tag{2.35}$$

Making operations gives the following:

$$\begin{bmatrix}f_1+1 & f_2 & f_3\end{bmatrix}=\begin{bmatrix}0 & 0 & 0\end{bmatrix}, \tag{2.36}$$

where the solution is given as follows:

$$\begin{aligned}f_1&=-1.0,\\ f_2&=0.0,\\ f_3&=0.0.\end{aligned} \tag{2.37}$$

So, F_s is found as follows:

$$F_s=\begin{bmatrix}-1 & 0 & 0\end{bmatrix}. \tag{2.38}$$

So, the structure regulator law is as follows:

$$\nu=\begin{bmatrix}-1 & 0 & 0\end{bmatrix}\chi. \tag{2.39}$$

2.5.3 Comparisons

In this section, the structure regulator of this chapter named structure is compared with the structure at infinity strategy of [8, 22] named structure at infinity. The benefit of the regulators is to attenuate the perturbations in the output of the plotter. The initial conditions are $\chi_0 = [1 \times 10^{-4}, 1 \times 10^{-4}, 0]^T$, it is desired that the plotter remains at rest even if its discs receive an undesired voltage. The perturbations are an undesired voltage given as $d = 0.25 \sin(20t)$ V for 0–4 s, t is the time in s.

The structure at infinity regulator of [8, 22] with the regulator law (2.32) is employed for the perturbation attenuation in the plotter (2.6 and 2.29).

The structure regulator of this study with the regulator law (2.27) of Theorem 2.2 is utilized for the perturbation attenuation in the plotter (2.6 and 2.29). The condition of Theorem 2.3 is satisfied with $O = \begin{bmatrix} 0.0025067 & 0.025392 & 0.26727 \\ 0.025392 & 103.14 & 1028.6 \\ 0.26727 & 1028.6 & 10849 \end{bmatrix}$, $U = \begin{bmatrix} 1 & 6.0704 \times 10^{-6} & 0.001445 \\ 6.0704 \times 10^{-6} & 0.99994 & -0.0036821 \\ 0.001445 & -0.0036821 & 0.88991 \end{bmatrix}$; so, the other values are $\overline{d}^2 = 0.25^2 = 0.0625$, $\alpha = \frac{1}{\theta} B_s^T O B_s \overline{d}^2 = \frac{1}{1 \times 10^{-2}}(2.5067 \times 10^{-3})(0.0625) = 1.5667 \times 10^{-2}$, $\gamma = \frac{\lambda_{\max}(O)}{\lambda_{\min}(O)} = \frac{10947}{2.5001 \times 10^{-3}} = 4.3786 \times 10^6$, $\beta = \lambda_{\min}(UO^{-1}) = 8.1234 \times 10^{-5}$, and $\xi = \frac{\alpha}{\beta \lambda_{\min}(O)} = \frac{1.5667 \times 10^{-2}}{8.1234 \times 10^{-5}(2.5001 \times 10^{-3})} = 77142$. The condition of Theorem 2.4 is satisfied with the eigenvalues of A_s are -1.1129×10^{-2}, -8.8726×10^{-2}, -200, where all the real parts are negative; so, the states of the regulator in the process will converge to $\lim_{t\to\infty} |\chi_1| \le 0.00125$, $\lim_{t\to\infty} |\chi_2| \le 0.00625$, and $\lim_{t\to\infty} |\chi_3| \le 0$.

Figures 2.8–2.11 show input and perturbations, states, output, and RMSE of the structure and structure at infinity applied for the perturbation attenuation in the plotter, respectively. Table 2.6 shows the RMSE of (2.15).

From Figures 2.8 and 2.9, it can be seen that in structure the regulator law is varying with the time, this fact is used for the perturbation attenuation and influences the movement of the states; while in structure at infinity the regulator law does not have a behavior. From Table 2.6 and Figures 2.10 and 2.11, it can be observed that structure achieves better accuracy when compared with structure at infinity for the perturbation attenuation, because in the first the output follows better the reference output than in the second, and because the RMSE is smaller for the first than for the second. So, Structure obtains its benefit, the perturbation attenuation in the output of the plotter.

Remark 2.7: From (2.27), the proposed strategy shows that the perturbations are not employed in the structure regulator law to be attenuated in the plotter.

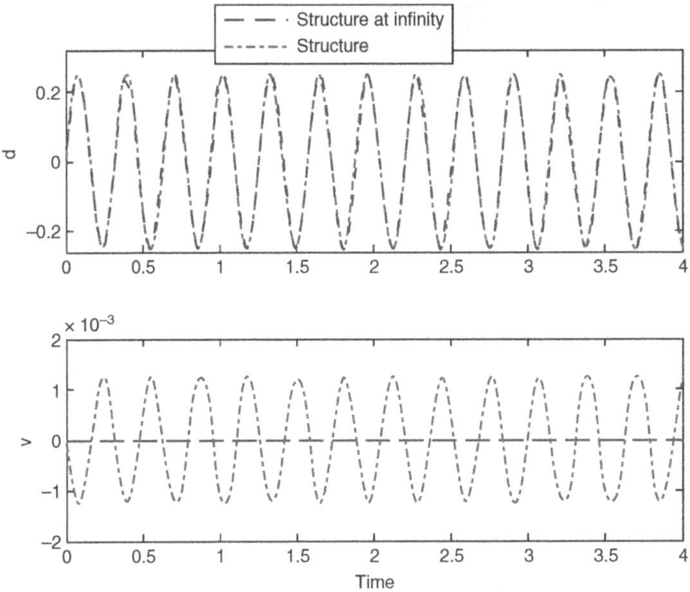

Figure 2.8 Input and perturbations of the plotter.

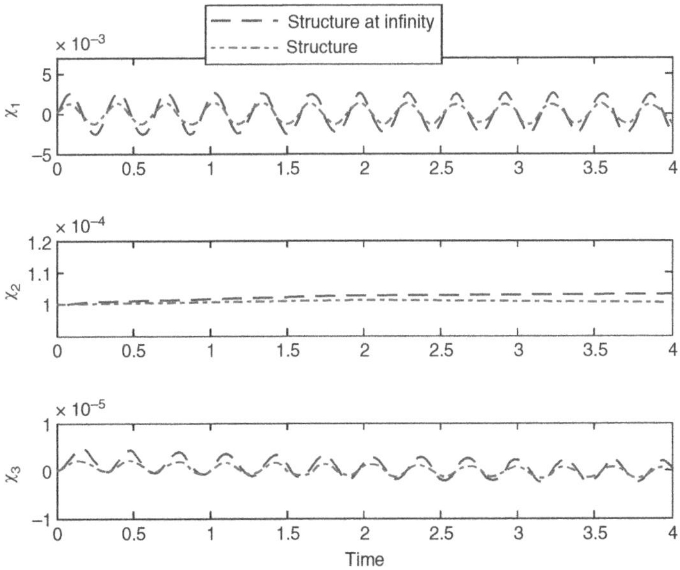

Figure 2.9 States of the plotter.

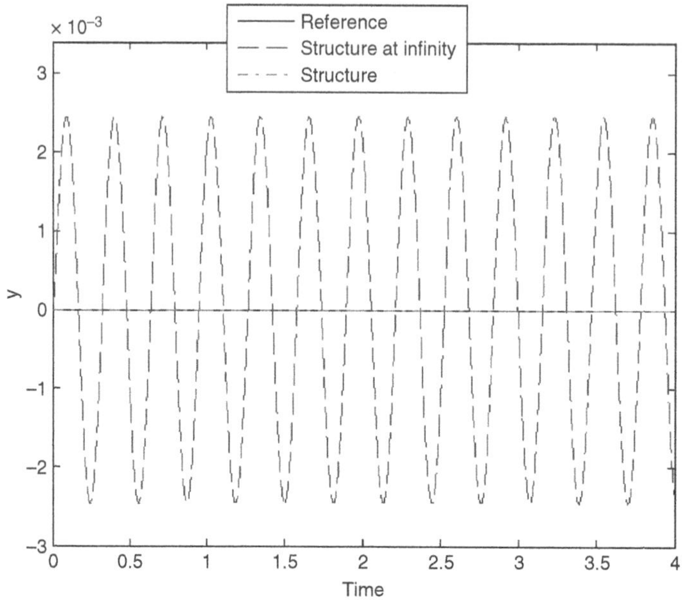

Figure 2.10 Output of the plotter.

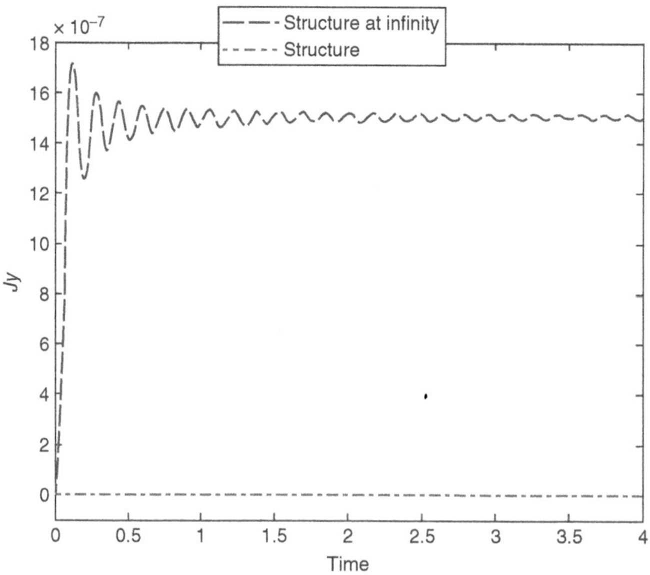

Figure 2.11 RMSE of the plotter.

Table 2.6 Comparisons for the plotter.

Strategies	Jy
Structure at infinity	1.4999×10^{-6}
Structure	0

2.6 Concluding Remarks

In this chapter, the structure regulator for the perturbation attenuation was designed, which acts in the inputs and affect the outputs of the processes. The strategy was based on the structure at infinity regulator. The stability of the regulator was warranted by the Lyapunov strategy; in addition, the convergence of the regulators was warranted. The structure regulator was compared with the structure at infinity strategy in a three-phase circuit and a plotter showed that the first regulator achieved better accuracy when compared with the second because the first obtained a better perturbation attenuation. The recommended strategy can be applied to many processes, as are the electric, mechanic, hydraulic, or thermal. In the future, other kinds of regulators will be designed for the perturbation attenuation, or the trajectory tracking problem will be considered.

References

1 Ahn C. K., Shi P., Basin M. V. (2015), Two-dimensional dissipative control and filtering for Roesser model, IEEE Transactions on Automatic Control, 60 (7), 1745–1759.

2 Ahn C. K., Shi P. (2015), Dissipativity analysis for fixed-point interfered digital filters, Signal Processing, 109, 148–153.

3 Ahn C. K. (2015), Some new results on the stability of direct-form digital filters with finite word length nonlinearities, Signal Processing, 108, 549–557.

4 Azhmyakov V., Basin M. V., Gil Garcia A. E. (2014), Optimal control processes associated with a class of discontinuous control systems: applications to sliding mode dynamics, Kybernetika, 50 (1), 5–18.

5 Guo B. Z., Zhou H. C., AL-Fhaid A. S., Younas A. M. M., Asiri A. (2015), Parameter estimation and stabilization for one-dimensional Schrödinger equation with boundary output constant disturbance and non-collocated control, Journal of the Franklin Institute, 352, 2047–2064.

6 Jiang X.-W., Hu B., Guan Z.-H., He D.-X., Zhang X.-H. (2015), Optimal tracking performance for non-square plant models with input disturbance and feedback channel noise, Journal of the Franklin Institute, 352, 2971–2984.

7 Lu Q., Zhang L., Basin M., Tian H. (2012), Analysis and synthesis for networked control systems with uncertain rate of packet losses, Journal of the Franklin Institute, 349, 2500–2514.

8 Rubio J. J., Figueroa M., Perez Cruz J. H., Rumbo J. (2012), Control to stabilize and mitigate disturbances in a rotary inverted pendulum, Revista Mexicana de Fisica E, 58 (2), 107–112.

9 Rubio J. J., Zamudio Z., Pacheco J., Mujica Vargas D., (2013), Proportional derivative control with inverse dead-zone for pendulum systems, Mathematical Problems in Engineering, 2013, 1–9.

10 Silva-Ortigoza R., Hernandez-Guzman V. M., Antonio-Cruz M., Munoz-Carrillo D. (2015), DC/DC buck power converter as a smooth starter for a DC motor based on a hierarchical control, IEEE Transactions on Power Electronics, 30 (2), 1076–1084.

11 Silva-Ortigoza R., Marquez-Sanchez C., Carrizosa-Corral F., Antonio-Cruz M., Alba-Martinez J. M., Saldaña-Gonzalez G. (2014), Hierarchical velocity control based on differential flatness for a DC/DC buck converter-DC motor system, Mathematical Problems in Engineering, 2014, 1–12.

12 Wei Y., Zheng W. X., Xu S. (2015), Robust output feedback control of uncertain time-delay systems with actuator saturation and disturbances, Journal of the Franklin Institute, 352, 2229–2248.

13 Xia Y., Dai L., Fu M., Li C., Wang C. (2014), Application of active disturbance rejection control in tank gun control system, Journal of the Franklin Institute, 351, 2299–2314.

14 Cheeger J., Tian G. (1994), On the cone structure at infinity of Ricci flat manifolds with euclidean volume growth and quadratic curvature decay, Inventiones Mathematicae, 118, 493–571.

15 Descusse J., Dion J. M. (1982), On the structure at infinity of linear square decoupled systems, IEEE Transactions on Automatic Control, AC-27 (4), 971–974.

16 Descusse J., Lafay J. F., Malabre M. (1982), On the structure at infinity of linear block-decouplable systems: the general clase, IEEE Transactions on Automatic Control, AC-28 (12), 1115–1118.

17 Laakkonen P., Pohjolainen S. (2011), Directed structure at infinity for infinite-dimensional systems, International Journal of Control, 84 (4), 702–715.

18 Longhi S., Perdon A., Conte G. (1989), Geometric and algebraic structure at infinity of discrete time linear periodic systems, Linear Algebra and its Applications, 122-124, 245–271.

19 Moog C. H. (1988), Nonlinear Decoupling and Structure at Infinity, Mathematics of Control, Signals and Systems, 1, 257–268.

20 Rabah R., Malabre M. (1993), Structure at infinity, model matching and disturbance rejection for linear systems with delays, Kybernetika, 29 (5), 485–498.

21 Rabah R., Malabre M. (1999), On the structure at infinity of linear delay systems with application to the disturbance decoupling problem, Kybernetika, 35 (6), 668–680.

22 Rubio J. J., Figueroa M., Perez-Cruz J. H., Bejarano F. J. (2012), Geometric approach and structure at infinite controls for the disturbance rejection, IET Control Theory and Applications, 6 (16), 2528–2537.

23 Sira-Raimrez H. (1990), Structure at infinity, zero dynamics and normal forms of systems undergoing sliding motions, International Journal of Systems Science, 21 (4), 665–674.

24 van der Wode J. W. (1991), On the structure at infinity of a structured system, Linear Algebra and Its Applications, 148, 145–169.

25 Vardulakis A. I. G., Karcanias N. (1983), Relations between strict equivalence invariants and structure at infinity of matrix pencils, IEEE Transactions on Automatic Control, AC-28 (4), 514–516.

26 Rubio J. J. (2016), Structure control for the disturbance rejection in two electromechanical processes, Journal of the Franklin Institute, 353 (14), 3610–3631.

27 Gracia-Calandin L. I., Perez-Vidal C. (2005), Modelado de sistemas dinamicos Applicaciones, Editorial Club Universitario, ISBN: 84-8454-422-2.

3

Hybrid Regulator with Observer for the Estimation and Attenuation of Perturbations

3.1 Introduction

Perturbations are undesired signals presented as inputs of a process which affect outputs. This problem has been studied in many processes. Attenuation of perturbations in the output of a process is required because it can affect the sensors, actuators, plants, or controls, causing accidents or unnecessary costs. So, a control to attenuate or to attenuate the perturbations in the output of a process is an important and current issue, in the theory, and applications.

There is some research about the perturbation attenuation or about regulators with observers. The criteria of robustness for a regulator to perturbations are evaluated in [1]. In [2], the optimal tracking performance of perturbed plants is considered. Various methods of improving the overall estimation quality in state observers are considered in [3]. In [4, 5], the sliding-mode approach is utilized for the perturbation attenuation. Methods for the perturbations detection are presented in [6, 7]. In [8], a novel control scheme based on perturbation attenuation is mentioned. An observer-based regulator using the output feedback is designed in [9]. In [10], an adaptive fuzzy backstepping output-feedback tracking control approach is recommended. An adaptive fuzzy robust output-feedback control problem is considered in [11]. In [12], a hybrid fuzzy adaptive output-feedback control design approach is recommended. A composite adaptive fuzzy output-feedback control approach is suggested in [13]. In [14], a fuzzy adaptive output-feedback stabilization control method is designed. A control for the stabilization is recommended in [15]. In [16], a fault-tolerant control of fuzzy processes is mentioned.

The above research mention some important investigations in regulators for the perturbation attenuation, such as [1–8], or in regulators with observers for the

Perturbations in Mechatronic, Robotic, and Energy Systems: Estimation and Attenuation, First Edition. Jose de Jesus Rubio. Published 2026 by John Wiley & Sons, Inc.

trajectory tracking such as [10–15]. However, it is not common to design regulators with observers for the estimation and attenuation of perturbations such as [9, 16].

In this chapter, a regulator with observer is suggested for the perturbation attenuation. The recommended strategy is divided into two designs: the observer, and the regulator with observer. The observer is selected to reach two main benefits: (1) to warrant its stability by the utilization of the Lyapunov strategy and (2) for the behavioral estimation of the unknown perturbations. The regulator with observer is selected to reach three main benefits: (1) to warrant its stability by the utilization of the Lyapunov strategy, (2) for the attenuation or attenuation of unknown perturbations, and (3) to suppose an alternative for the decreasing of chattering in the sliding-mode behavior.

On the other hand, the aforementioned research shows that the perturbation attenuation problem is an actual and novel investigation. In this chapter, an alternative method for the estimation and attenuation of perturbations is designed. The method is based on the combination of the sliding-mode strategy and the output-feedback strategy.

The sliding-mode regulator (SMR) has two problems: (1) the produced chattering because it can damage devices, and (2) the sliding-mode surface because it requires many derivatives of the process. A proposal of solution for both problems is presented in this research as follows: (1) the chattering of the SMR is decreased by utilizing the estimated perturbations from an observer, and (2) derivatives of the process are not required because the sliding-mode surface of this regulator is changed by the output-feedback strategy.

And finally, there are some works about stable perturbation attenuation. In [17], the perturbation attenuation of closed loop control processes is focused. Robust controls of turbines with perturbations are recommended in [18, 19], and [20]. In [21], the stabilization for a Schrodinger equation with perturbations is designed. The stability using perturbation attenuation control is approached in [22]. In [23], a robust regulator of perturbed processes is designed. The sliding-mode observer for jump processes against actuator faults is investigated in [24]. In [25], the perturbation attenuation control for stable processes is investigated. The H-infinity control of a neural network with an external perturbations is designed in [26]. In [27], the robust output-feedback control of processes with perturbations is explained. The robust stability of a class of perturbed processes is investigated in [28].

From the above studies, in most of the cases authors use the Lyapunov method. So, in this investigation, the Lyapunov analysis is utilized to warrant the exponential stability of the recommended regulator with observer.

This chapter is organized as follows. Section 3.2 presents the employed processes, they will be used for the regulator design. In Section 3.3, the hybrid observer is recommended for the perturbation estimation. In Section 3.4, the hybrid regulator with observer is recommended for the perturbation attenuation.

In Sections 3.5 and 3.6, the recommended strategy is verified by two examples: the plotter and suspension processes, respectively. Section 3.7 presents conclusions and suggests future research directions.

3.2 The Perturbed Process

Consider the following perturbed process:

$$\begin{aligned} \dot{\chi} &= A\chi + B\nu + Bd, \\ y &= C\chi, \end{aligned} \tag{3.1}$$

where $\chi \in \mathfrak{R}^n$ is the state, $\nu \in \mathfrak{R}^m$ is the control input, $y \in \mathfrak{R}^p$ is the output, $d \in \mathfrak{R}^m$ is the perturbations. $A \in \mathfrak{R}^{n\times n}$, $B \in \mathfrak{R}^{n\times m}$, and $C \in \mathfrak{R}^{p\times n}$ are matrices representing the transformations $\mathbf{A} : \mathfrak{R}^n \rightarrow \mathfrak{R}^n$, $\mathbf{B} : \mathfrak{R}^m \rightarrow \mathfrak{R}^n$, and $\mathbf{C} : \mathfrak{R}^n \rightarrow \mathfrak{R}^p$, respectively.

The following two sections will describe the observer for the estimation of perturbations, and the regulator with observer for the attenuation or attenuation of perturbations.

3.3 Hybrid Observer

3.3.1 Observer Design

In this section, a hybrid observer will be designed based on the assumption that the output y is known, and the perturbations d is unknown. Let $\widehat{\chi} \in \mathfrak{R}^n$ be the estimation of states $\chi \in \mathfrak{R}^n$. Define the output error $\widetilde{y}$ as follows:

$$\widetilde{y} = y - \widehat{y} = C\widetilde{\chi}, \tag{3.2}$$

where $y = C\chi \in \mathfrak{R}^p$, $\widehat{y} = C\widehat{\chi} \in \mathfrak{R}^p$, $\widetilde{\chi} = \chi - \widehat{\chi} \in \mathfrak{R}^n$ is the state error.

The following hybrid observer is recommended:

$$\dot{\widehat{\chi}} = A\widehat{\chi} + K\widetilde{y} + B\nu + Nsgn(M\widetilde{y}), \tag{3.3}$$

where $\widetilde{y}$ is the output error, $N \in \mathfrak{R}$ is a constant matrix which is selected after, M is a vector or matrix such that $MC \in \mathfrak{R}^{n\times n}$ is a positive semi-definite constant, K is a vector or matrix such that $KC \in \mathfrak{R}^{n\times n}$ is a constant matrix, sgn$(\cdot)$ is the signum function, M, N, and K, will be utilized in the stability analysis of the next section.

Remark 3.1: The observer of (3.2) and (3.3) is selected to find two main benefits: (1) to warrant its stability by the utilization of the Lyapunov strategy in the application of processes with the form (3.1) and (2) for the behavioral estimation of the unknown perturbations in the process (3.1).

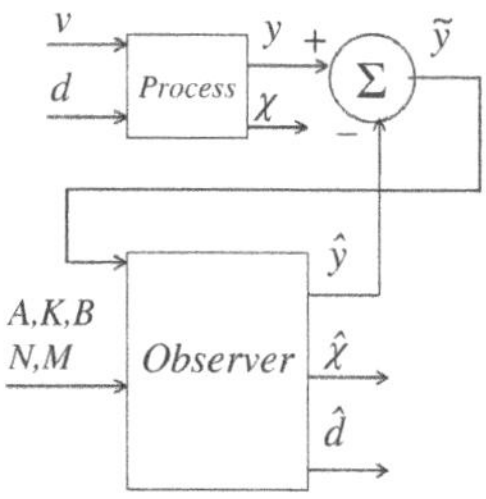

Figure 3.1 Hybrid observer.

The benefit of the observer is that states $\widehat{\chi}$ and the perturbations $\widehat{d}$ of (3.3) may follow states χ and the perturbations d of (3.1) using outputs y in the output error (3.2). See Figure 3.1.

3.3.2 Stability Analysis

In this section, the exponential stability of the closed loop in the hybrid observer and the perturbed process is warranted by the solution of the Lyapunov method.

From (3.1–3.3), the observer error equation can be formed as follows:

$$\begin{aligned}\dot{\widetilde{\chi}} &= A\widetilde{\chi} + Bd - K\widetilde{y} - Nsgn(M\widetilde{y})\\ &= [A - KC]\widetilde{\chi} + Bd - Nsgn(M\widetilde{y})\\ &= A_o\widetilde{\chi} + Bd - Nsgn(M\widetilde{y}),\end{aligned} \tag{3.4}$$

where $A_o = A - KC$.

Theorem 3.1 shows the stability of the recommended hybrid observer.

Theorem 3.1: The state of the hybrid observer (3.2) and (3.3) is applied to estimate the state χ of the perturbed process (3.1) and it warrants exponential stability; so, the observer error $\widetilde{\chi}$ satisfies:

$$\|\widetilde{\chi}\|^2 \leq \alpha e^{-\gamma t}\|\widetilde{\chi}_i\|^2, \tag{3.5}$$

where $\widetilde{\chi}_i$ is the initial condition of $\widetilde{\chi}$, $\alpha = \frac{\lambda_{\max}(S)}{\lambda_{\min}(S)}$, $\gamma = \lambda_{\min}(RS^{-1})$, $\|d\| \leq \overline{d}$, $\overline{d}$ is the known upper bound of d, d are the unknown perturbations given in (3.1), $\|Bd\| \leq \|B\|\,\overline{d} \leq N$, $MC \geq 0$, $\|\cdot\|$ is the Euclidean norm in $\Re^n$ and $|\cdot|$ is the absolute value, $S \in \Re^{n\times n}$ and $R \in \Re^{n\times n}$ are positive matrices which satisfy the following equation:

$$A_o^T S + SA_o = -R, \tag{3.6}$$

where A_o is given in (3.4).

Proof. See [29] for the proof. ■

Remark 3.2: The Routh–Hurwitz criterion is related with the recommended stability in that: (1) if (3.6) has a solution, then eigenvalues of A_o are in the left half plane, (2) if observer error $\tilde{\chi}$ satisfies (3.5), then the process with a characteristic equation is stable. However, the benefit of the theorem is (3.5), which describes the form of the convergence in the observer error. It starts from an initial value and has an exponential convergence to zero.

3.3.3 Estimation of the Perturbations

The result of (3.5) describes that the state $\|\tilde{\chi}\|^2$ starts from the initial value $\|\tilde{\chi}_i\|^2$ and will converge with an exponential behavior to zero. So, the following fact is true:

$$\lim_{t\to\infty} \|\tilde{\chi}\|^2 = 0, \tag{3.7}$$

so $\lim_{t\to\infty} \tilde{\chi} = 0$ and $\lim_{t\to\infty} \dot{\tilde{\chi}} = 0$, substituting it in (3.4) given as follows:

$$\begin{aligned} &\lim_{t\to\infty} \dot{\tilde{\chi}} = \lim_{t\to\infty} A\tilde{\chi} + \lim_{t\to\infty} Bd - \lim_{t\to\infty} K\tilde{y} - \lim_{t\to\infty} Nsgn(M\tilde{y}) \\ &\Longrightarrow 0 = \lim_{t\to\infty} Bd - \lim_{t\to\infty} K\tilde{y} - \lim_{t\to\infty} Nsgn(M\tilde{y}) \\ &\Longrightarrow \lim_{t\to\infty} d = \lim_{t\to\infty} B^+K\tilde{y} + \lim_{t\to\infty} B^+Nsgn(M\tilde{y}), \end{aligned} \tag{3.8}$$

where B^+ is the pseudo-inverse of B. As all the elements of (3.8) are bounded, the perturbations can be estimated as follows:

$$\hat{d} = B^+K\tilde{y} + B^+Nsgn(M\tilde{y}), \tag{3.9}$$

where $\hat{d}$ are the estimation of the perturbations d.

Remark 3.3: Note that the pseudo-inverse is correctly applied in this research, this concept is more deeply studied in [30].

3.4 Hybrid Regulator with Observer

3.4.1 Regulator Design

In this section, a hybrid regulator with observer will be designed based on the assumption that the output y is known, and the perturbations d are unknown. Let $\hat{\chi} \in \mathfrak{R}^n$ be the estimation of states $\chi \in \mathfrak{R}^n$. Let $\hat{d} \in \mathfrak{R}^m$ of (3.9) be the estimation of the perturbations $d \in \mathfrak{R}^m$ of (3.1). Define the perturbations error $\tilde{d}$ as follows:

$$\tilde{d} = d - \hat{d}, \tag{3.10}$$

The following hybrid regulator with observer is recommended:

$$\nu = -\hat{d} - B^{+}Ky - B^{+}Nsgn(My), \tag{3.11}$$

where ν is the control input, y is the output, $\hat{d}$ are the estimation of the perturbations d, $N \in \Re$ is a constant matrix which is selected after, M is a vector or matrix such that $MC \in \Re^{nxn}$ is a positive semi-definite constant, K is a vector or matrix such that $KC \in \Re^{nxn}$ is a constant matrix, sgn$(\cdot)$ is the signum function, B^{+} is the pseudo-inverse of B. Note that N, M, and K are the same parameters used in the observer (3.3), M, N, and K, will be utilized in the stability analysis of the next section.

Remark 3.4: The regulator with observer of (3.2, 3.3, and 3.11), is selected to find three main benefits: (1) to warrant its stability by the utilization of the Lyapunov strategy in the application of processes with the form (3.1), (2) for the attenuation or attenuation of the unknown perturbations in the process (3.1), and (3) to suppose an alternative for decreasing of the chattering in the sliding-mode behavior.

The benefit of the hybrid regulator with observer is that the output y should converge to zero while the perturbations $\hat{d}$ of (3.9) and (3.11) may follow the perturbations d of (3.1). See Figure 3.2.

Remark 3.5: Please note that the recommended regulator does not require the knowledge of the perturbations behavior because it uses the estimated perturbations in control law.

3.4.2 Stability Analysis

In this section, the exponential stability of the closed loop in the hybrid regulator with observer and the perturbed process is warranted by the solution of the Lyapunov method.

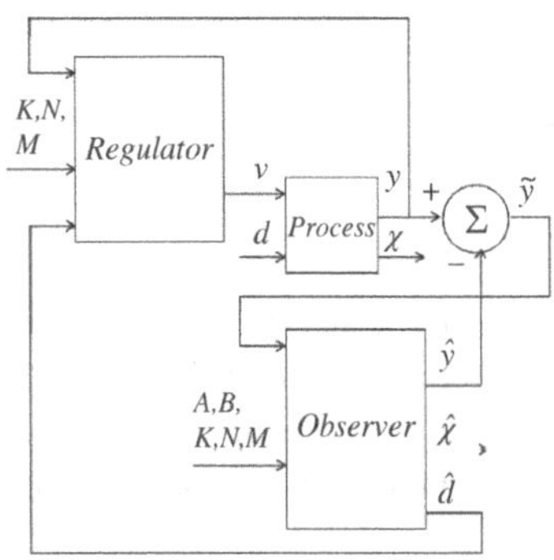

Figure 3.2 Hybrid regulator.

From (3.1), (3.10), and (3.11), the regulator with observer error equation can be formed as follows:

$$\begin{aligned} \dot{\chi} &= A\chi + B\left[d - \hat{d}\right] - Ky - Nsgn(My) \\ &= [A - KC]\,\chi + B\tilde{d} - Nsgn(My) \\ &= A_o\chi + B\tilde{d} - Nsgn(My), \end{aligned} \tag{3.12}$$

where $A_o = A - KC$. Note that A_o is the same parameter used in the observer (3.4), $\tilde{d}$ is given in (3.10).

The following theorem shows the stability of the recommended hybrid regulator with observer.

Theorem 3.2: The state of the hybrid regulator with observer (3.2, 3.3, 3.9, and 3.11), is applied to attenuate the perturbations of the perturbed process (3.1) and it warrants exponential stability; so, the state χ satisfies:

$$\|\chi\|^2 \leq \alpha e^{-\gamma t}\|\chi_i\|^2, \tag{3.13}$$

where χ_i is the initial condition of χ, $\alpha = \frac{\lambda_{\max}(S)}{\lambda_{\min}(S)}$, $\gamma = \lambda_{\min}(RS^{-1})$, $\overline{d}$ is the known upper bound of $\tilde{d}$, $\tilde{d}$ is the unknown perturbations error given in (3.10), $\left\|B\tilde{d}\right\| \leq \|B\|\,\overline{d} \leq N$, $MC \geq 0$, $\|\cdot\|$ is the Euclidean norm in $\mathfrak{R}^n$ and $|\cdot|$ is the absolute value, $S \in \mathfrak{R}^{nxn}$ and $R \in \mathfrak{R}^{nxn}$ are positive matrices which satisfy the following equation:

$$A_o^T S + SA_o = -R, \tag{3.14}$$

where A_o is given in (3.4 and 3.12).

Proof. See [29] for the proof. ■

Remark 3.6: Considering the utilization of the observer to estimate the perturbations, if there is a good estimation of the perturbations, then the perturbations error norm $\tilde{d}$ of the regulator with observer of (3.13) is smaller than the perturbations norm d of the regulator of [4], it is named $\left\|\tilde{d}\right\| \leq \|d\|$. So, in this special case, it can be seen that $\left\|B\tilde{d}\right\|$ of (3.13) in the regulator with observer is smaller than $\|Bd\|$ in the regulator of [4]. And finally, in this special case, N in the regulator with observer of (3.2, 3.3, 3.9, and 3.11), must be smaller than N in the regulator of [4], decreasing chattering effects in the recommended strategy.

Remark 3.7: Please note that the recommended regulator does not require the knowledge of the perturbations behavior, it only requires the knowledge of its upper bound in the stability analysis.

Remark 3.8: The Routh–Hurwitz criterion is related with the recommended stability in that: 1) if (3.14) has a solution, then the eigenvalues of A_o are in the left half plane, 2) if state χ satisfies (3.13), then the process with a characteristic equation is stable. However, the benefit of the theorem is (3.13) which describes the form of the convergence in the state, it starts from an initial value and has an exponential convergence to zero.

In the following two sections, the hybrid regulator with observer of (3.2, 3.3, 3.9, and 3.11), named HR will be compared with the SMR of [4] named SMR and with the observer-based regulator of [9] named R for the perturbation attenuation in two examples. The root mean square error (RMSE) for states will be used, it is given as follows:

$$\text{RMSE} = \left(\frac{1}{T} \int_0^T x^2 d\tau \right)^{\frac{1}{2}}, \tag{3.15}$$

where $x^2 = \sum_{i=1}^{n} \chi_i^2$ for states, $x^2 = y^2$ for the output, or $x^2 = \left(d - \hat{d}\right)$ for the perturbations error.

3.5 Plotter Process

Figure 3.3 shows the plotter. The plotter is a computer printer for printing vector graphics. It has a DC motor, a disc 1, a disc 2, a flywheel, a disc 3, and a poppet [31].

The states are $\chi_1 = i$ (electric current in the motor in A), $\chi_2 = \theta_1$ (angle of the disc 1 in rad), $\chi_3 = \omega_1$ (velocity of the disc 1 in rad/s), the input is $\nu = V_e$

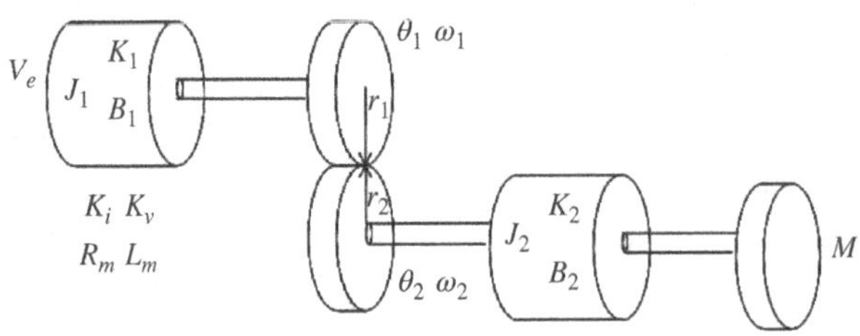

Figure 3.3 Plotter process.

(input voltage in V), and the output is $y = R_m i = R_m \chi_1$, the mathematical model is as follows:

$$\begin{bmatrix} \dot{\chi}_1 \\ \dot{\chi}_2 \\ \dot{\chi}_3 \end{bmatrix} = A \begin{bmatrix} \chi_1 \\ \chi_2 \\ \chi_3 \end{bmatrix} + Bv + Bd,$$

$$y = C \begin{bmatrix} \chi_1 \\ \chi_2 \\ \chi_3 \end{bmatrix}, \tag{3.16}$$

where:

$$A = \begin{bmatrix} -\frac{R_m}{L_m} & 0 & -\frac{K_v}{L_m} \\ 0 & 0 & 1 \\ \frac{K_i}{J_1 + \frac{r_1^2}{r_2^2}(J_2 + M)} & -\frac{K_1 + \frac{r_1^2}{r_2^2}K_2}{J_1 + \frac{r_1^2}{r_2^2}(J_2 + M)} & -\frac{B_1 + \frac{r_1^2}{r_2^2}B_2}{J_1 + \frac{r_1^2}{r_2^2}(J_2 + M)} \end{bmatrix},$$

$$B = \begin{bmatrix} \frac{1}{L_m} & 0 & 0 \end{bmatrix}^T,$$

$$C = \begin{bmatrix} R_m & 0 & 0 \end{bmatrix},$$

and:

R_m is the resistance of the motor, which is; 1 Ω
L_m = inductance of the motor = 10×10^{-3} H
J_1 = inertia of the disc 1 = 0.01213 kgm^2
J_2 = inertia of the disc 2 = 0.01213 kgm^2
B_1 = viscous friction of the disc 1 = 0.1 kgm^2rad/s
B_2 = viscous friction of the disc 2 = 0.1 kgm^2rad/s
M = mass of the pen nib = 0.5 kg
K_i = motor magnetic flux of the motor = 1×10^{-3} kgm^2/s^2
K_v = back emf of the motor = 0.45 Vs/rad
K_1 = spring effect of the disc 1 = 1×10^{-3} kgm^2/s^2
K_2 = spring effect of the disc 2 = 1×10^{-3} kgm^2/s^2
r_1 = radius of the disc 1 = 0.1 m
r_2 = radius of the disc 2 = 0.1 m

Using the values of above parameters, matrices of the process (3.16) are rewritten as follows:

$$A = \begin{bmatrix} -100 & 0 & -45.0 \\ 0 & 0 & 1 \\ 4.9372 \times 10^{-3} & -9.8743 \times 10^{-4} & -9.8743 \times 10^{-2} \end{bmatrix},$$

$$B = \begin{bmatrix} 100 & 0 & 0 \end{bmatrix}^T,$$

$$C = \begin{bmatrix} 1 & 0 & 0 \end{bmatrix},$$

In this example, the mathematical model of the plotter is given by (3.1) and (3.16), 0.1, 0.1, 0.1 are considered as initial conditions for plant states χ_1, χ_2, χ_3. 0, 0, 0 are used as the initial states for the observers $\widehat{\chi}_1$, $\widehat{\chi}_2$, $\widehat{\chi}_3$. The sine perturbations are $d = 0.2\sin(20t)$ from 0 s to 4 s for a first case, and the square perturbations are $d = 0.2square(20t)$ from 0 s to 4 s for a second case.

The benefit of the SMR is that the plant output y of (3.1) and (3.16) should converge to zero with the regulator of [4]. The SMR is given with parameters

$$B^+ = \begin{bmatrix} \frac{1}{100} & 0 & 0 \end{bmatrix}, K = \begin{bmatrix} 1 \times 10^3 \\ 1 \times 10^3 \\ 0 \end{bmatrix}, M = \begin{bmatrix} 1 \\ 1 \\ 1 \end{bmatrix}, N = 10(0.2).$$

The benefit of the observer-based regulator (R) is that the plant output y of (3.1) and (3.16) should converge to zero with the regulator of [9], while the perturbations $\widehat{d}$ of [9] may follow the disturbance d of (3.1). The R is given with parameters

$$B^+ = \begin{bmatrix} \frac{1}{100} & 0 & 0 \end{bmatrix}, K = \begin{bmatrix} 1 \times 10^3 \\ 1 \times 10^3 \\ 0 \end{bmatrix}.$$

The benefit of the hybrid regulator (HR) with observer is that the plant output y of (3.1) and (3.16) should converge to zero with the regulator of (3.11), while the perturbations $\widehat{d}$ of (3.9) may follow the perturbations d of (3.1). The HR is given by (3.3, 3.9, and 3.11), with parameters $B^+ = \begin{bmatrix} \frac{1}{100} & 0 & 0 \end{bmatrix}$, $K = \begin{bmatrix} 1 \times 10^3 \\ 1 \times 10^3 \\ 0 \end{bmatrix}$, $M = \begin{bmatrix} 1 \\ 1 \\ 1 \end{bmatrix}$, $N = 0.2$. Lyapunov equations (3.6) and (3.14) are satisfied with $S = \begin{bmatrix} 5.1621 & -5.6801 & -459.53 \\ -5.6801 & 6.251 & 506.35 \\ -459.53 & 506.35 & 2.1455 \times 10^5 \end{bmatrix}$,

$$R = \begin{bmatrix} 0.95758 & -6.3705 \times 10^{-2} & 0.32297 \\ -6.3705 \times 10^{-2} & 0.99997 & -3.8755 \times 10^{-3} \\ 0.32297 & -3.8755 \times 10^{-3} & 0.2213 \end{bmatrix}.$$

Considering the Theorems 3.1 and 3.2, states of the hybrid regulator with observer are exponentially stable with parameters $\alpha = 1.5615 \times 10^{-5}$ and $\gamma = 5.1681 \times 10^{8}$.

Figures 3.4–3.7 show the perturbations, input, states, and output for regulators in the sine perturbations. Table 3.1 shows the RMSE for regulators using (3.15) for the sine perturbations.

Note that the SMR requires bigger gains than the HR to obtain satisfactory results. It could cause an incorrect functioning of the SMR because it could yield the undesired saturation of its gains. Note that in Table 3.1, SMR does not have a value in the RMSE for d, it is because this regulator does not use an observer and d is not estimated. From Figures 3.4 to 3.7, it can be seen that the HR improves both the SMR and R because the perturbations of the first follow the plant perturbations better than the third, while the second cannot estimate the perturbations, and because the output of the first converges faster to zero than for the other. From Table 3.1, it can be shown that the HR achieves better accuracy when compared

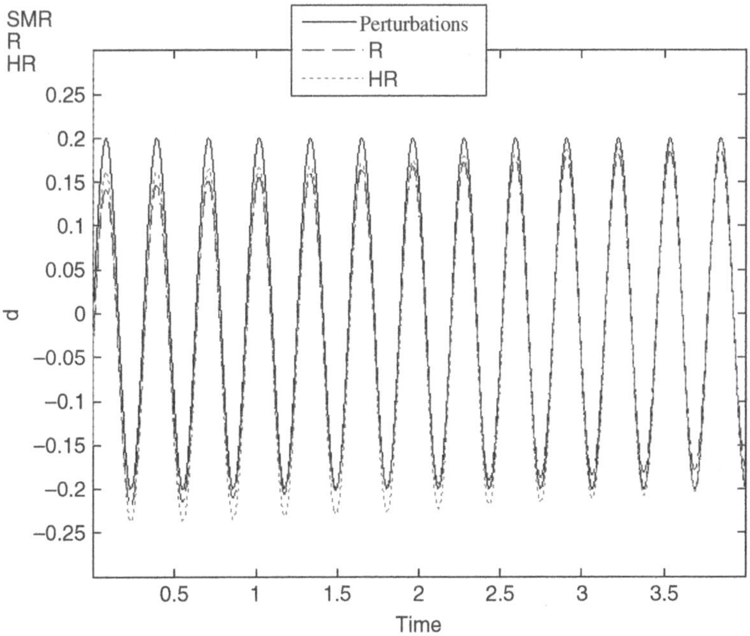

Figure 3.4 Perturbations for the plotter process.

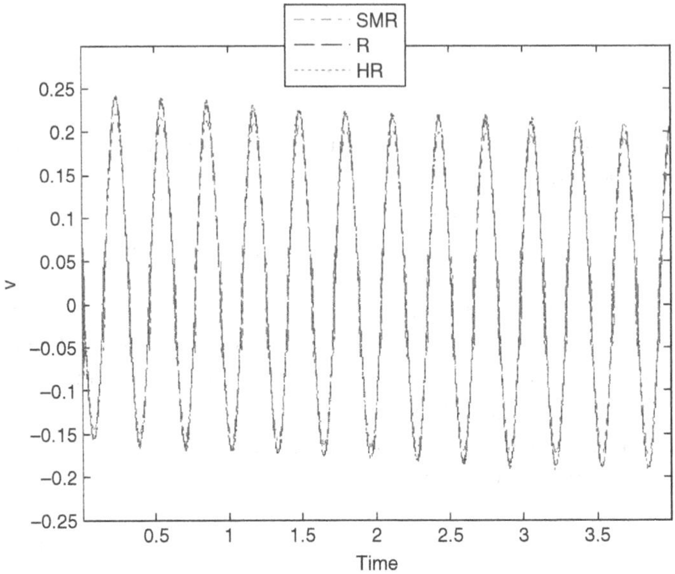

Figure 3.5 Input for the plotter process.

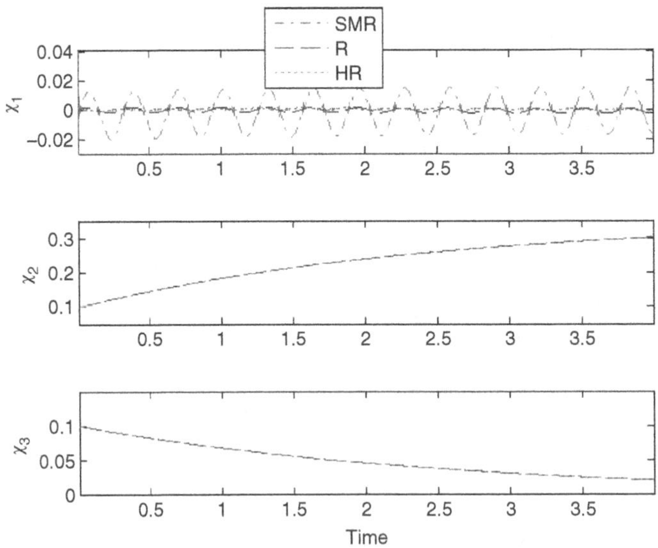

Figure 3.6 States for the plotter process.

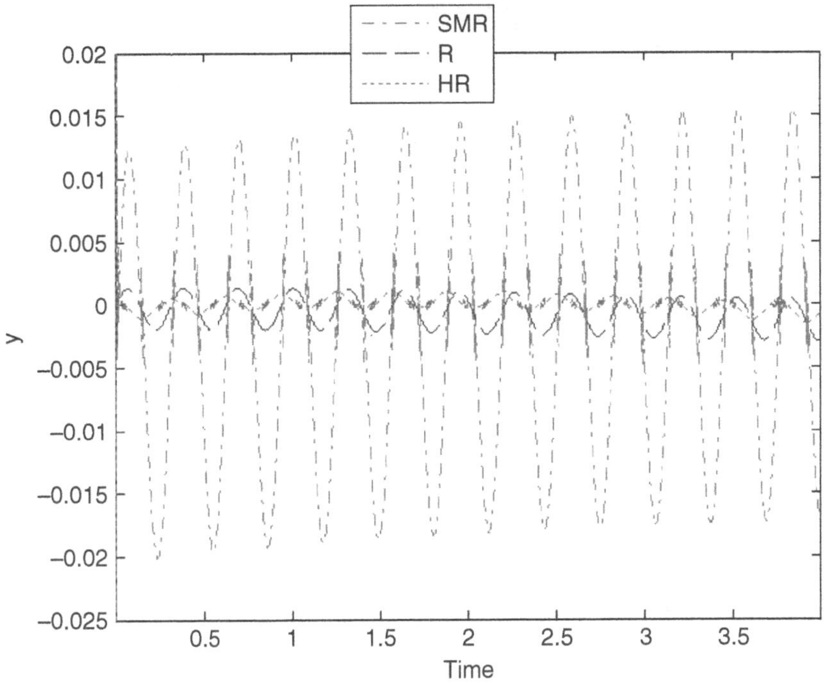

Figure 3.7 Output for the plotter process.

Table 3.1 The RMSE for regulators.

Strategies	RMSE for d	RMSE for χ	RMSE for y
SMR	–	0.1698	0.0081
R	0.0221	0.1697	0.0020
HR	0.0219	0.1696	0.0018

with both the SMR and R because the RMSE for the first is smaller than for the other.

Figures 3.8–3.11 show the perturbations, input, states, and output for regulators in the square perturbations. Table 3.2 shows the RMSE for regulators using (3.15) for the square perturbations.

Note that the SMR requires bigger gains than the HR to obtain satisfactory results. It could cause an incorrect functioning of the SMR because it could yield the undesired saturation of its gains. Note that in Table 3.2, SMR does not have a

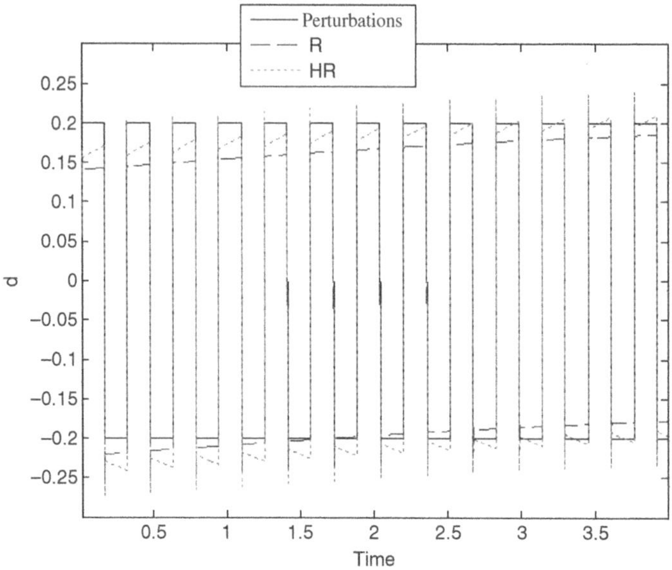

Figure 3.8 Perturbations for the plotter process.

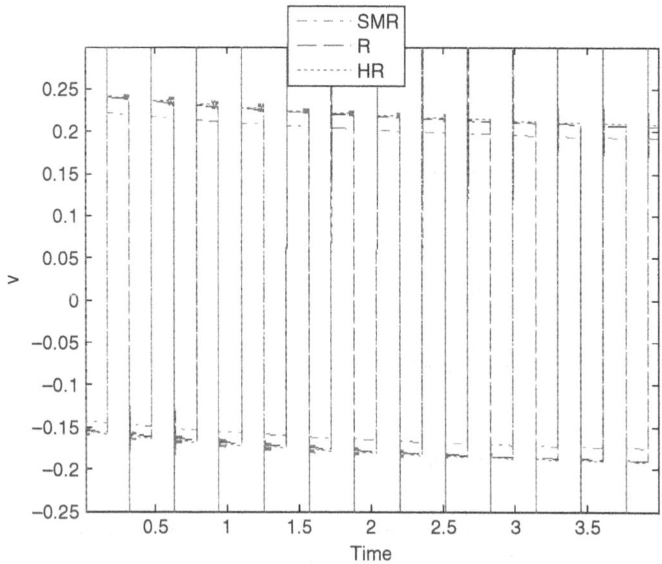

Figure 3.9 Input for the plotter process.

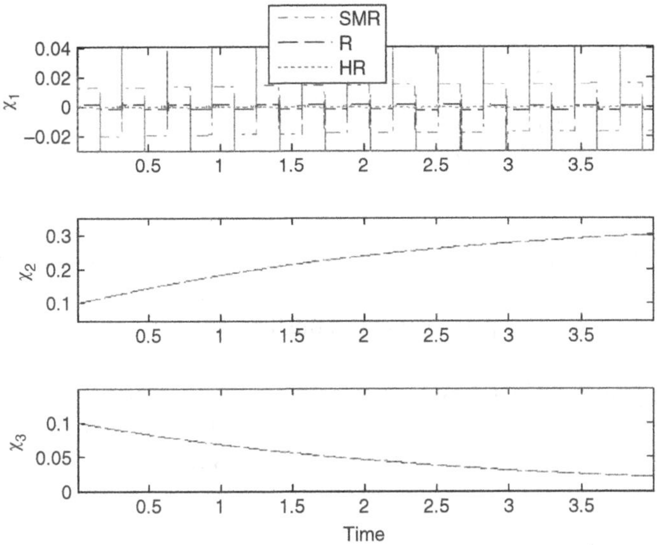

Figure 3.10 States for the plotter process.

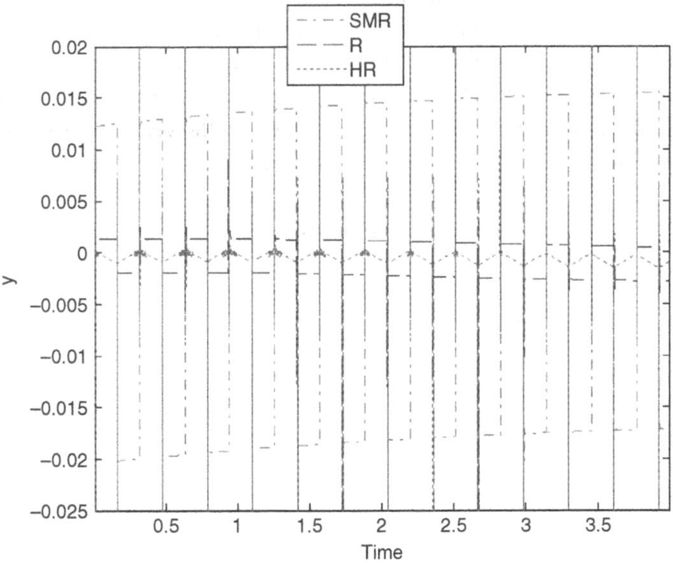

Figure 3.11 Output for the plotter process.

Table 3.2 The RMSE for regulators.

Strategies	RMSE for d	RMSE for χ	RMSE for y
SMR	–	0.1700	0.0117
R	0.0218	0.1697	0.0030
HR	0.0175	0.1696	0.0028

value in the RMSE for d, it is because this regulator does not use an observer and d is not estimated. From Figures 3.8–3.11, it can be seen that the HR improves both the SMR and R because the perturbations of the first follow the plant perturbations better than the third, while the second cannot estimate the perturbations, and because the output of the first converges faster to zero than for the other. From Table 3.2, it can be shown that the HR achieves better accuracy when compared with both the SMR and R because the RMSE for the first is smaller than for the other.

3.6 Suspension Process

Figure 3.12 shows a diagram of the partial suspension process in a vehicle. The process consists of a servo-valve that activates a piston, which is connected with the main mass of the vehicle m and with the wheel-tire set. The variable to be controlled is the displacement of this set, because the irregularities of the road create displacements that generate forces, which must be absorbed by the piston to exert a null strain on the vehicle. x is the displacement of the vehicle mass, y is the displacement of the wheel, and u is the main displacement that the road exerts on the wheel.

The model of the process in state space is as follows:

$$\begin{bmatrix} \dot{\chi}_1 \\ \dot{\chi}_2 \\ \dot{\chi}_3 \end{bmatrix} = A \begin{bmatrix} \chi_1 \\ \chi_2 \\ \chi_3 \end{bmatrix} + Bv + Bd,$$
$$y = C \begin{bmatrix} \chi_1 \\ \chi_2 \\ \chi_3 \end{bmatrix}, \tag{3.17}$$

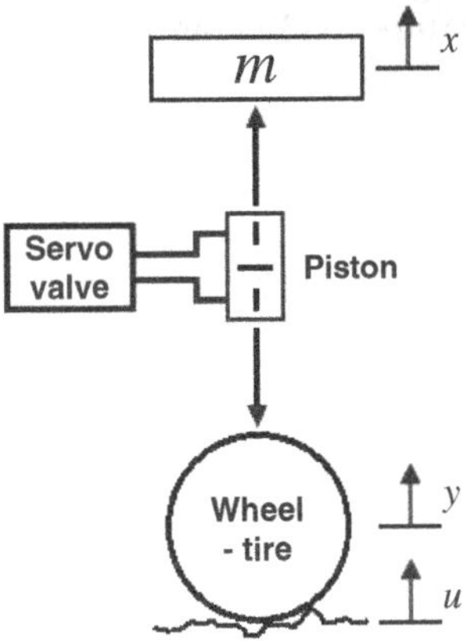

Figure 3.12 Suspension process.

where

$$A = \begin{bmatrix} 0 & 1 & 0 \\ 0 & 0 & -\frac{k_t}{m} \\ G_a k_i k_1 & 1 + G_a k_i k_2 & -\frac{k_t}{B_i} \end{bmatrix},$$

$$B = \begin{bmatrix} 0 \\ \frac{k_t}{m} \\ \frac{k_t}{B_i} + G_a k_i k_1 \end{bmatrix},$$

$$C = \begin{bmatrix} 0 & 0 & 1 \end{bmatrix},$$

where

k_t = tire stiffness = 2.8×10^5 N/m
B_i = cross-line damping coefficient = 5930 N s/m
m = mass of the vehicle = 240 kg
G_a = servo-amplifier gain = 50 mA/V
k_i = servo-valve flow coefficient = 5.27×10^{-8}
k_1 = feedback gain 1 = 2.44×10^8
k_2 = feedback gain 2 = -1.876×10^8
State variables are defined as follows:
$\chi_1 = x$ = displacement of the mass in the vehicle
$\chi_2 = \dot{x}$ = speed of the mass in the vehicle
$\chi_3 = y$ = displacement of the wheel
$\nu = u$ = displacement that the road exerts on the wheel.

Using the values of above parameters, matrices of the process (3.17) are rewritten as follows:

$$A = \begin{bmatrix} 0 & 1 & 0 \\ 0 & 0 & -1166.667 \\ 0.64294 & 0.50567 & -47.217 \end{bmatrix},$$

$$B = \begin{bmatrix} 0 \\ 1166.667 \\ 47.86 \end{bmatrix},$$

$$C = \begin{bmatrix} 0 & 0 & 1 \end{bmatrix}.$$

In this example, the mathematical model of the suspension process is given by (3.1) and (3.17), 0.1, 0.1, 0.1 are considered as initial conditions for plant states χ_1, χ_2, χ_3. 0, 0, 0 are used as initial states for observers $\widehat{\chi}_1$, $\widehat{\chi}_2$, $\widehat{\chi}_3$. The sine perturbations are $d = 0.2\sin(20t)$ from 0 s to 4 s for a first case, and the ramp perturbations are $d = 0.2\,ramp(20t)$ from 0 s to 4 s for a second case.

The benefit of the SMR is that the plant output y of (3.1) and (3.17) should converge to zero with the regulator of [4]. The SMR is given with parameters $B^+ = \begin{bmatrix} 0 & \frac{0.5}{1166.667} & \frac{0.5}{47.86} \end{bmatrix}$, $K = \begin{bmatrix} 1\times10^4 \\ 1\times10^4 \\ 1\times10^3 \end{bmatrix}$, $M = \begin{bmatrix} 1 \\ 1 \\ 1 \end{bmatrix}$, $N = 10(0.2)$.

The benefit of the observer-based regulator (R) is that the plant output y of (3.1) and (3.17) should converge to zero with the regulator of [9], while the perturbations $\widehat{d}$ of [9] may follow the perturbations d of (3.1). The R is given with parameters $B^+ = \begin{bmatrix} 0 & \frac{0.5}{1166.667} & \frac{0.5}{47.86} \end{bmatrix}$, $K = \begin{bmatrix} 1\times10^4 \\ 1\times10^4 \\ 1\times10^3 \end{bmatrix}$.

The benefit of the hybrid regulator with observer (HR) is that the plant output y of (3.1) and (3.17) should converge to zero with the regulator of (3.11), while the perturbations $\widehat{d}$ of (3.9) may follow the perturbations d of (3.1). The HR is given by (3.3, 3.9, and 3.11) with parameters $B^+ = \begin{bmatrix} 0 & \frac{0.5}{1166.667} & \frac{0.5}{47.86} \end{bmatrix}$, $K = \begin{bmatrix} 1\times10^4 \\ 1\times10^4 \\ 1\times10^3 \end{bmatrix}$, $M = \begin{bmatrix} 1 \\ 1 \\ 1 \end{bmatrix}$, $N = 0.2$. Lyapunov equations

(3.6 and 3.14) are satisfied with $S = \begin{bmatrix} 0.52292 & -0.39802 & -0.77767 \\ -0.39802 & 0.37488 & -0.20169 \\ -0.77767 & -0.20169 & 10.024 \end{bmatrix}$,

$$R = \begin{bmatrix} 0.99999 & -1.0425 \times 10^{-6} & 6.9944 \times 10^{-3} \\ -1.0425 \times 10^{-6} & 1.0 & 2.9489 \times 10^{-3} \\ 6.9944 \times 10^{-3} & 2.9489 \times 10^{-3} & 0.07954 \end{bmatrix}.$$

Considering the Theorems 3.1 and 3.2, the states of the hybrid regulator with observer are exponentially stable with parameters $\alpha = 7.9129 \times 10^{-3}$ and $\gamma = 20242$.

Figures 3.13–3.16 show the perturbations, input, states, and output for regulators in the sine perturbations. Table 3.3 shows the RMSE for regulators using (3.15) for the sine perturbations.

Note that the SMR requires bigger gains than the HR to obtain satisfactory results. It could cause an incorrect functioning of the SMR because it could yield the undesired saturation of its gains. Note that in Table 3.3, SMR does not have a value in the RMSE for d, it is because this regulator does not use an observer

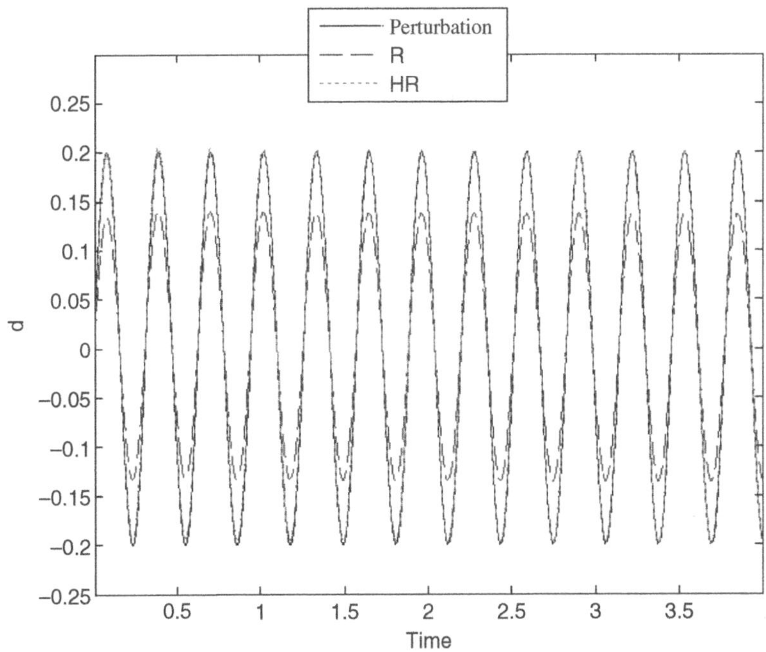

Figure 3.13 Perturbation for the suspension process.

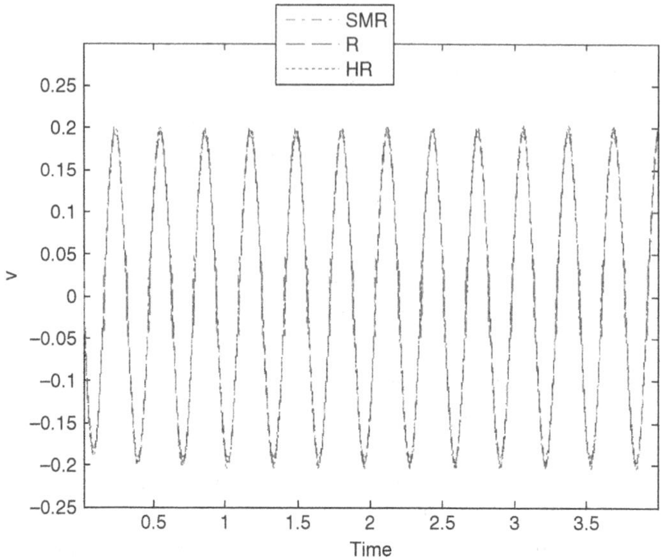

Figure 3.14 Input for the suspension process.

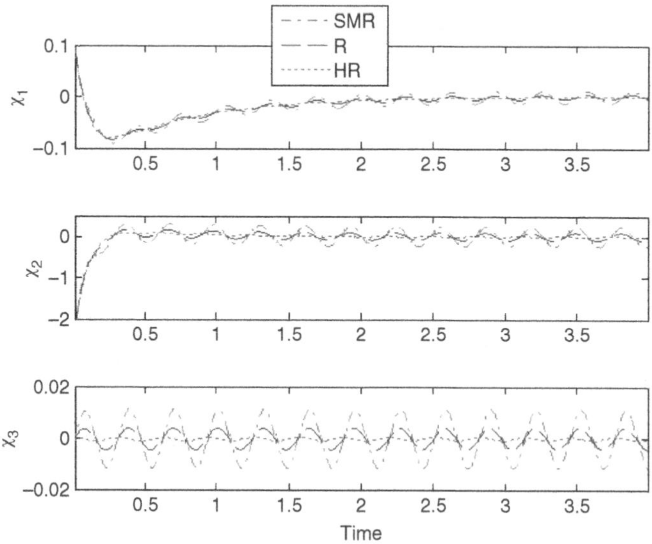

Figure 3.15 States for the suspension process.

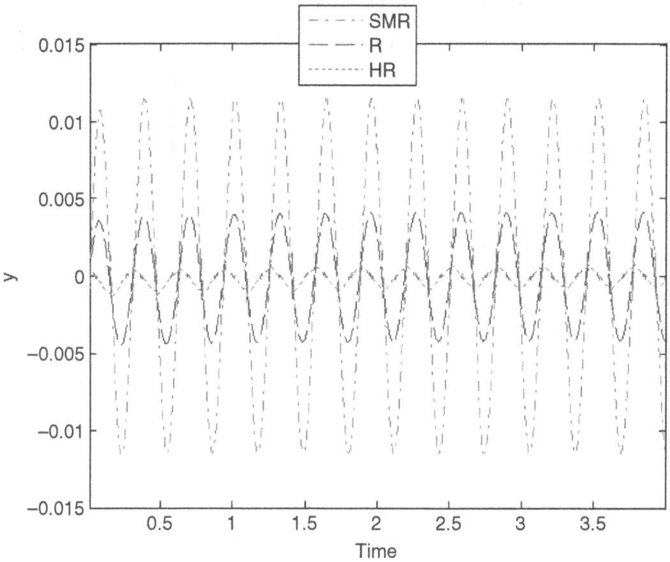

Figure 3.16 Output for the suspension process.

Table 3.3 The RMSE for regulators.

Strategies	RMSE for d	RMSE for χ	RMSE for y
SMR	–	0.2026	0.0057
R	0.0361	0.1762	0.0024
HR	0.0244	0.1750	0.0015

and d is not estimated. From Figures 3.13 to 3.16, it can be seen that the HR improves both the SMR and R because the perturbations of the first follow better than the plant perturbations of the third, while the second cannot estimate the perturbations, and because the output of the first converge faster to zero than for the other. From Table 3.3, it can be shown that the HR achieves better accuracy when compared with both the SMR and R because the RMSE for the first is smaller than for the other.

Figures 3.17–3.20 show the perturbations, input, states, and output for regulators in the ramp perturbations. Table 3.4 shows the RMSE for regulators using (3.15) for the ramp perturbations.

Note that the SMR requires bigger gains than the HR to obtain satisfactory results. It could cause an incorrect functioning of the SMR because it could yield the undesired saturation of its gains. Note that in Table 3.4, SMR does not have a

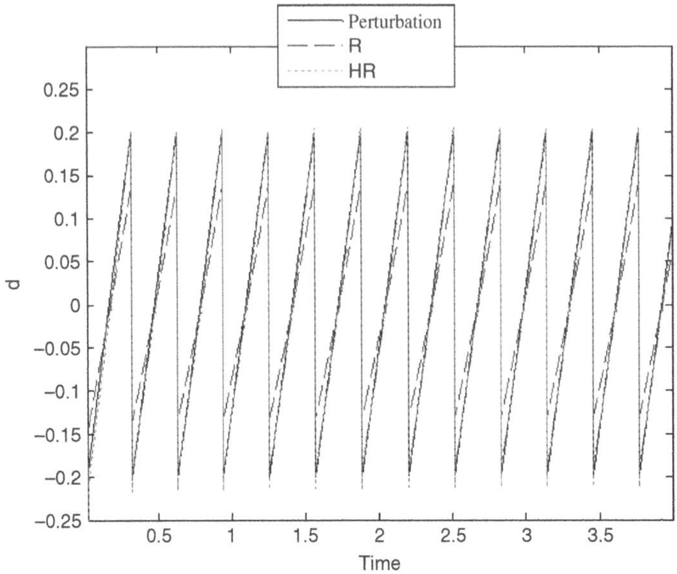

Figure 3.17 Perturbation for the suspension process.

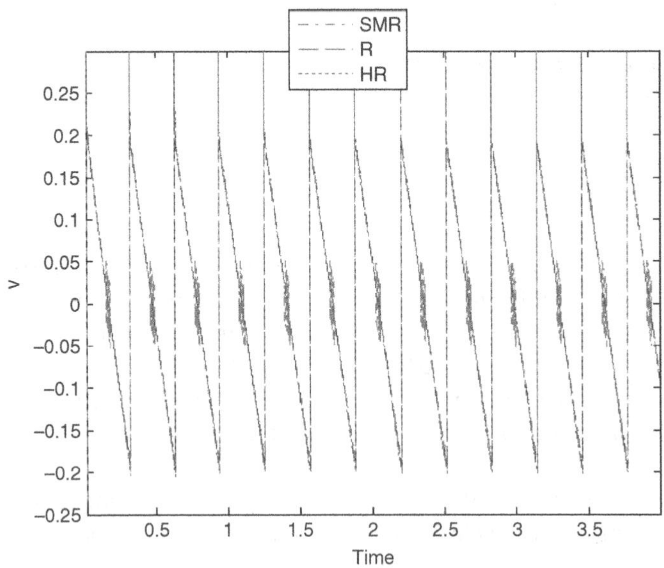

Figure 3.18 Input for the suspension process.

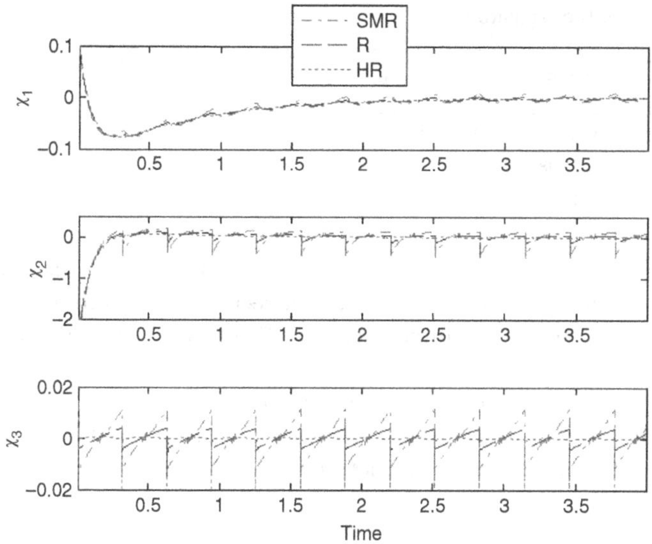

Figure 3.19 States for the suspension process.

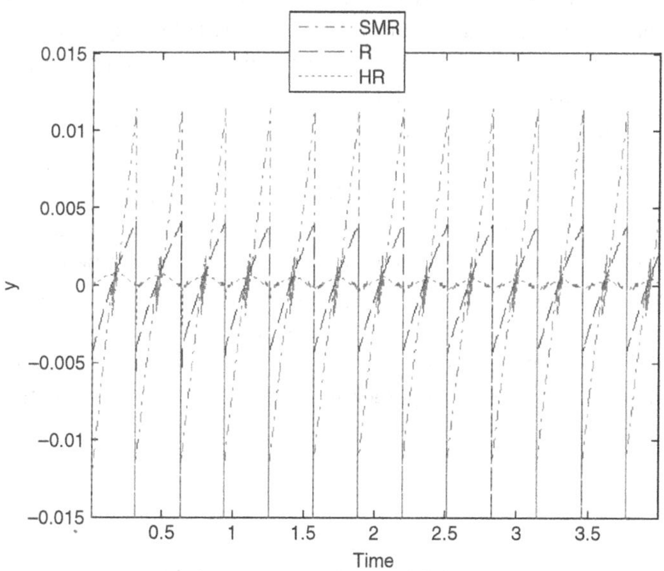

Figure 3.20 Output for the suspension process.

Table 3.4 The RMSE for the regulators.

Strategies	RMSE for d	RMSE for χ	RMSE for y
SMR	–	0.2061	0.0045
R	0.0319	0.1808	0.0022
HR	0.0270	0.1755	0.0017

value in the RMSE for d, it is because this regulator does not use an observer and d is not estimated. From Figures 3.17 to 3.20, it can be seen that the HR improves both the SMR and R because the perturbations of the first follow better than the plant perturbations of the third, while the second cannot estimate the perturbations, and because the output of the first converges faster to zero than for the other. From Table 3.4, it can be shown that the HR achieves better accuracy when compared with both the SMR and R because the RMSE for the first is smaller than for the other.

3.7 Concluding Remarks

In this chapter, the usefulness of the hybrid approach is shown for solving the perturbation estimation problem; furthermore, the usefulness of the hybrid strategy is shown for solving the perturbation attenuation problem. Results showed that the hybrid regulator with observer achieves better accuracy when compared with both the SMR and the observer-based regulator for the estimation and attenuation of the sine, square, and ramp perturbations in a plotter and a suspension process. This chapter represents the theoretical basis for developing perturbation estimation problems via an observer, and for developing perturbation attenuation problems via a regulator, for being able afterward to perform an implementation in physical processes. The recommended strategy can be applied to many mechatronic processes such as electric, mechanic, hydraulic, or thermal. In the future, other kinds of regulators or observers will be designed for the attenuation or estimation of perturbations, or the trajectory tracking problem of these kinds of processes will be considered.

References

1 Hernandez O., Ortiz P., Herrera J. (2015), Fractional PID controller with LQR proportional action applied to fractional model of cement rotary kiln, IEEE Latin America Transactions, 13 (1), 37–42.

2 Jiang X. W., Hua B., Guan Z. H., He D. X., Zhang X. H. (2015), Optimal tracking performance for non-square plant models with input disturbance and feedback channel noise, Journal of the Franklin Institute, 352, 2971–2984.

3 Madonski R., Herman P. (2015), Survey on methods of increasing the efficiency of extended state disturbance observers, ISA Transactions, 56, 18–27.
4 Rubio J. J., Pacheco J., Juarez C. F., Soriano E. (2016), Comparison between two observers, IEEE Latin America Transactions, 14 (5), 2077–2084.
5 Rubio J. J. (2016), Structure control for the disturbance rejection in two electromechanical processes, Journal of the Franklin Institute, 353 (14), 3610–3631.
6 Sayed-Mouchaweh M., Lughofer E. (2015), Decentralized fault diagnosis approach without a global model for fault diagnosis of discrete event systems, International Journal of Control, 88 (11), 2228–2241.
7 Toubakh H., Sayed-Mouchaweh M. (2016), Hybrid dynamic classifier for drift-like fault diagnosis in a class of hybrid dynamic systems: application to wind turbine converters, Neurocomputing, 171, 1496–1516.
8 Shao X., Wang H. (2015), Active disturbance rejection based trajectory linearization control for hypersonic reentry vehicle with bounded uncertainties, ISA Transactions, 54, 27–38.
9 Rubio J. J., Ochoa G., Balcazar R., Pacheco J. (2015), Uniform stable observer for the disturbance estimation in two renewable energy systems, ISA Transactions, 58, 155–164.
10 Li Y., Tong S., Li T. (2015), Observer-based adaptive fuzzy tracking control of MIMO stochastic nonlinear systems with unknown control directions and unknown dead zones, IEEE Transactions on Fuzzy Systems, 23 (4), 1228–1241.
11 Li Y., Tong S., Liu Y., Li T. (2014), Adaptive fuzzy robust output feedback control of nonlinear systems with unknown dead zones based on a small-gain approach, IEEE Transactions on Fuzzy Systems, 22 (1), 164–176.
12 Li Y., Tong S., Li T. (2015), Composite adaptive fuzzy output feedback control design for uncertain nonlinear strict-feedback systems with input saturation, IEEE Transactions on Cybernetics, 45 (10), 2299–2308.
13 Li Y., Tong S., Li T. (2016), Hybrid fuzzy adaptive output feedback control design for uncertain mimo nonlinear systems with time-varying delays and input saturation, IEEE Transactions on Fuzzy Systems, 24 (4), 841–853.
14 Li Y., Tong S. (2017), Adaptive fuzzy output-feedback stabilization control for a class of switched nonstrict-feedback nonlinear systems, IEEE Transactions on Cybernetics, 47 (4), 1007–1016.
15 Aguilar-Ibanez C. (2009), The Lyapunov direct method for the stabilisation of the ball on the actuated beam, International Journal of Control, 82 (12), 2169–2178.
16 Huang S. J., Yang G. H. (2014), Fault tolerant controller design for T-S fuzzy systems with time-varying delay and actuator faults: a k-step fault-estimation approach, IEEE Transactions on Fuzzy Systems, 22 (6), 1526–1540.
17 Alagoz B. B., Deniz F. N., Keles C., Tan N. (2015), Disturbance rejection performance analyses of closed loop control systems by reference to disturbance ratio, ISA Transactions, 55, 63–71.

18 Ginter V. J., Pieper J. K. (2011), Robust gain scheduled control of a hydrokinetic turbine, IEEE Transactions on Control Systems Technology, 19 (1), 805–817.

19 Jafarnejadsani H., Pieper J. (2015), Gain-scheduled l1-optimal control of variable-speed-variable-pitch wind turbines, IEEE Transactions on Control Systems Technology, 23 (1), 372–379.

20 Jafarnejadsani H., Pieper J., Ehlers J. (2013), Adaptive control of a variable-speed variable-pitch wind turbine using radial-basis function neural network, IEEE Transactions on Control Systems Technology, 21 (6), 2264–2272.

21 Guo B. Z., Zhou H. C., AL-Fhaid A. S., Younas A. M. M., Asiri A. (2015), Parameter estimation and stabilization for one-dimensional Schrodinger equation with boundary output constant disturbance and non-collocated control, Journal of the Franklin Institute, 352, 2047–2064.

22 Li J., Qi X., Xia Y., Pu F., Chang K. (2015), Frequency domain stability analysis of nonlinear active disturbance rejection control system, ISA Transactions, 56, 188–195.

23 Niu B., Zhao X., Zhang L., Li H. (2015), p-times differentiable unbounded functions for robust control of uncertain switched nonlinear systems with tracking constraints, International Journal of Robust and Nonlinear Control, 25, 2965–2983.

24 Shi P., Liu M., Zhang L. (2015), Fault-tolerant sliding-mode-observer synthesis of Markovian jump systems using quantized measurements, IEEE Transactions on Industrial Electronics, 62 (9), 5910–5918.

25 Wang L., Su J. (2015), Disturbance rejection control for non-minimum phase systems with optimal disturbance observer, ISA Transactions, 57, 1–9.

26 Wang Z., Huang L., Guo Z. (2015), $H\infty$ control for neural networks with discontinuous activations and nonlinear external disturbance, Journal of the Franklin Institute, 352, 3144–3165.

27 Wei Y., Zheng W. X., Xu S. (2015), Robust output feedback control of uncertain time-delay systems with actuator saturation and disturbances, Journal of the Franklin Institute, 352, 2229–2248.

28 Zhao X., Zhang L., Shi P., Reza Karimi H. (2014), Robust control of continuous-time systems with state-dependent uncertainties and its application to electronic circuits, IEEE Transactions on Industrial Electronics, 61 (8), 4161–4170.

29 Rubio J. J. (2016), Hybrid controller with observer for the estimation and rejection of disturbances, ISA Transactions, 65, 445–455.

30 Rubio J. J., Aquino V., Figueroa M. (2013), Inverse kinematics of a mobile robot, Neural Computing and Applications, 23 (1), 187–194.

31 Gracia-Calandin L. I., Perez-Vidal C. (2005), Modelado de sistemas dinamicos Applicaciones, Editorial Club Universitario, ISBN: 84-8454-422-2.

4

Sliding-mode Regulator for the Perturbations Attenuation in Two Tank Processes

4.1 Introduction

The perturbation attenuation in the process output and states has been recently studied. The perturbation attenuation is required since the perturbations can affect the sensors, actuators, processes, or regulators, causing accidents or unnecessary costs. Since a regulator is employed to get that all the process states reach constant behaviors, a regulator for the perturbation attenuation is an important and current topic in the theory and applications.

There are some works about the stable regulation for the perturbation attenuation. In [1–4], strategies for the perturbations detection in processes are addressed. The geometric approach for the perturbation attenuation in perturbed processes is studied in [5, 6]. In [7–10], active perturbation attenuation regulators of processes with perturbations are recommended. Sliding-mode regulators (SMRs) for processes with perturbations are recommended in [11–13]. In [14–17], stable regulators for the perturbation attenuation are focused. H-infinity regulators of perturbed processes are addressed in [18–22]. In [23–26], several kinds of regulators for perturbed processes are discussed. The aforementioned development shows that a regulator for the perturbation attenuation is a novel topic.

From the above research, in [1, 2, 4, 7, 8, 10, 11, 13, 14, 16, 22, 24], several kinds of observers are employed for the perturbation estimation, and these estimated perturbations are taken into account in the mentioned regulators as the perturbations compensation. It must be desired to design a regulator for the perturbation attenuation in which an observer is not required because it must make the development more compact.

In this chapter, a strategy is explained for the perturbation attenuation in the process output and states. For this purpose, the next three benefits in the recommended regulator are taken into account:

1. Contrary with the interesting and mentioned works, the recommended regulator does not employ an observer for the perturbation estimation. In this regulator, the sliding-mode behavior is employed as the perturbations compensation. The benefit of employing a regulator and not an observer-based regulator, is that it makes the recommended strategy more compact.
2. The SMR has an issue in the sliding mode surface because it sometimes requires many derivatives of the process [27, 28]. The difference of this study with respect to the previous strategies is that the recommended regulator does not employ the sliding mode surface. It also makes the recommended strategy more compact.
3. The Lyapunov strategy is employed to warrant the stability of the SMR for the perturbation attenuation. The recommended approach only requires the boundedness of perturbations.

This chapter is organized as follows. Section 4.2 explains the perturbed processes, they will be used for the regulator design. The SMR for the perturbation attenuation in the process output and states is recommended in Section 4.3. In Sections 4.4 and 4.5, the SMR is employed in two tank processes. Section 4.6 describes conclusions and suggests future development directions.

4.2 The Perturbed Process

Take into account the next perturbed process:

$$\begin{aligned} \dot{\chi} &= A\chi + B\nu + Bp, \\ y &= C\chi, \end{aligned} \tag{4.1}$$

where $\chi \in \Re^n$ are the states, $\nu \in \Re^m$ are the regulator inputs, $y \in \Re$ is the output, $p \in \Re^m$ are the perturbations. $A \in \Re^{n\times n}$, $B \in \Re^{n\times m}$, and $C \in \Re^{1\times n}$ are matrices representing the maps $\mathbf{A} : \Re^n \to \Re^n$, $\mathbf{B} : \Re^m \to \Re^n$, and $\mathbf{C} : \Re^n \to \Re$, respectively.

The reference model is:

$$\begin{aligned} \dot{\chi}_r &= A\chi_r, \\ y_r &= C\chi_r, \end{aligned} \tag{4.2}$$

where $\chi_r \in \Re^n$ are the reference states and $y_r \in \Re$ is the reference output.

The output error is:

$$\begin{aligned} \widetilde{y} &= y - y_r \\ &= C\widetilde{\chi} = C(\chi - \chi_r), \end{aligned} \tag{4.3}$$

where $\widetilde{y} \in \Re$ is the output error and $\widetilde{\chi} \in \Re^n$ are the states errors.

4.3 The Sliding Mode Regulator

4.3.1 The Regulator Development

In this section, a regulator will be developed based on the assumption that the output y is measured, but the perturbations p are not measured.

The next SMR is recommended:

$$\nu = -B^* L\widetilde{y} - B^* K sgn(\widetilde{y}), \tag{4.4}$$

where $\nu \in \Re^m$ are the regulator inputs, $\widetilde{y} \in \Re$ is the output error, $K \in \Re^n$ is a constant vector which is chosen after, $L \in \Re^n$ is a constant vector, $sgn(\cdot) \in \Re$ is the signum map, $B^* \in \Re^{m\times n}$ is the pseudo-inverse of $B \in \Re^{n\times m}$.

The purpose of the regulator (4.4) is that the output y of the perturbed process (4.1) should converge to the reference output y_r of the reference model (4.2) employing the output y. See Figure 4.1.

Remark 4.1: Note that the behaviors of the perturbations p are not measured in the described regulator because it only requires the measure of their upper bound.

4.3.2 The Stability Analysis

In this section, states of the SMR applied to the perturbed processes are warranted to be stable by using the solution of the Lyapunov strategy.

Subtracting the reference model (4.2) to the model (4.1) and employing the states errors (4.3):

$$\begin{aligned} &\dot{\chi} - \dot{\chi}_r = A(\chi - \chi_r) + B\nu + Bp \\ &\Rightarrow \dot{\widetilde{\chi}} = A\widetilde{\chi} + B\nu + Bp. \end{aligned} \tag{4.5}$$

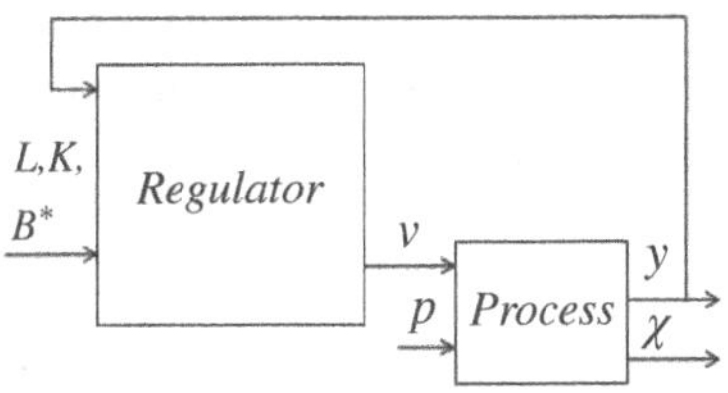

Figure 4.1 The sliding-mode regulator.

Substituting the SMR (4.4) into (4.5) and employing the output error (4.3) forms the closed-loop regulator:

$$\begin{aligned}\dot{\tilde{\chi}} &= A\tilde{\chi} + Bp - L\tilde{y} - Ksgn(\tilde{y}) \\ &= [A - LC]\tilde{\chi} + Bp - Ksgn(\tilde{y}) \\ &= A_{smr}\tilde{\chi} + Bp - Ksgn(\tilde{y}),\end{aligned} \tag{4.6}$$

where $A_{smr} = A - LC$, p is described in (4.1).

Theorem 4.1 shows the stability of the recommended SMR.

Theorem 4.1: States of the SMR (4.4), used for the perturbation attenuation in the process (4.1) are asymptotic stable; so, states errors $\tilde{\chi}$ comply:

$$\|\tilde{\chi}\|^2 \leq \rho e^{-\sigma t}\|\tilde{\chi}_i\|^2, \tag{4.7}$$

where $\tilde{\chi}_i$ are the initial conditions of $\tilde{\chi}$, $\rho = \frac{\lambda_{\max}(T)}{\lambda_{\min}(T)}$, $\sigma = \lambda_{\min}(WT^{-1})$, $|Bp| \leq B\overline{p}$ and $\frac{B\overline{p}}{sgn(C)} \leq K$, p are the perturbations described in (4.1), $\|\cdot\|$ is the Euclidean norm in $\mathfrak{R}^n$ and $|\cdot|$ is the absolute value, $T \in \mathfrak{R}^{n\times n}$ and $W \in \mathfrak{R}^{n\times n}$ are positive matrices which comply the next equality:

$$A_{smr}^T T + TA_{smr} = -W, \tag{4.8}$$

where A_{smr} is described in (4.6).

Proof. See [29] for the proof. ■

Remark 4.2: The solvability of the mentioned regulator is that the solution of (4.8) gives the gain L which is employed in the regulator (4.4) to reach the perturbation attenuation in the perturbed process (4.1).

In Sections 4.4 and 4.5, the SMR of (4.4) named SMR will be compared with the geometric regulator of [6] named GR in two experiments. The root mean square error (MSE) for the output and states will be exploited, it is:

$$\text{MSE} = \left(\frac{1}{T}\int_0^T \chi^2 d\tau\right)^{\frac{1}{2}}, \tag{4.9}$$

where $\chi^2 = \sum_{j=1}^{n}\tilde{\chi}_j^2$ for the states errors, or $\chi^2 = \tilde{y}^2$ for the output error.

4.4 The Thermal Process

This process consists of three interconnected tanks [6]. It is understood that the liquid enters the tank with a fixed temperature; it sends the tank liquid to a determined temperature. The valves allow a discharge of liquid outside the process. The purpose is to keep the liquid temperature in the tanks at a desired value.

Figure 4.2 shows the thermal process which is employed to get the real data of perturbations and to validate a model. Angular keys with a barrel type of 1/2 inch are adapted as valves. Servo motors of 0.1 revolutions per minute are employed to move the valves. An aluminium resistance of 200 W and 120 V is employed to heat the water. Temperature sensors LM35 are employed to get the temperature measures. An Arduino microcontroller is employed to send the temperature measures to a personal computer. The personal computer is employed to receive and to save the temperature measures.

Figure 4.3 shows the thermal process diagram, where t_{02}, t_{03} are the temperatures of the water, which fall in the tanks 1, 3 (°C), t_1, t_2, t_3 are the temperatures of the water for the tanks 1, 2, 3 (°C), f_1, f_2, f_3 are the flows of the water for

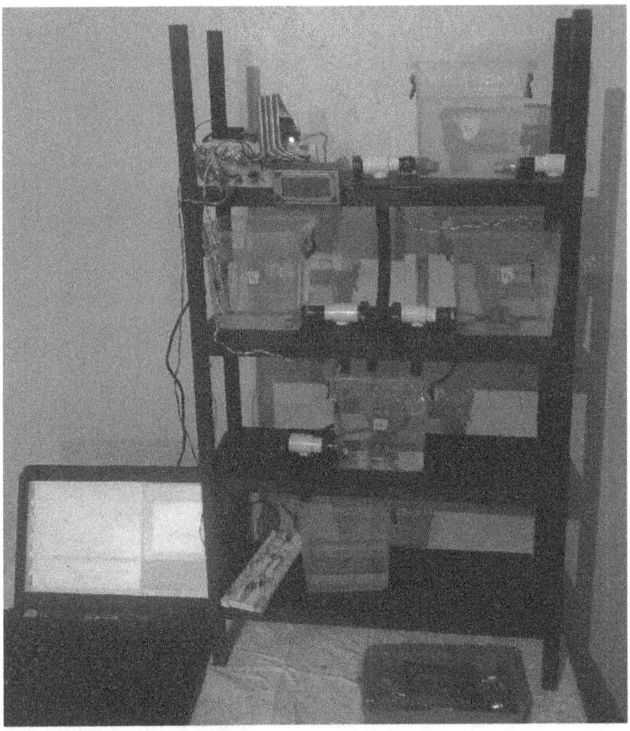

Figure 4.2 The thermal process. *Source:* Rubio et al. [29].

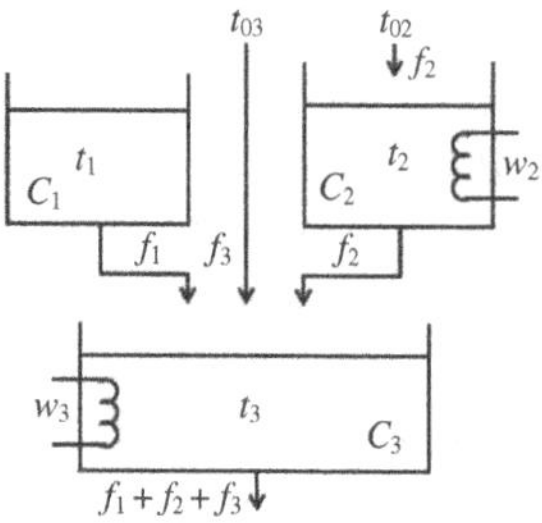

Figure 4.3 The thermal process diagram.

the tanks 2, 3 (m^3/s), C_1, C_2, C_3 are the calorific capacities of the water for the tanks 1, 2, 3 ($kcal/m^3 °C$), and w_1, w_2 are the heating energy employed in the tanks 2, 3 ($kcal/m^3$).

The model of the thermal process in state space is:

$$\begin{bmatrix} \dot{x}_1 \\ \dot{x}_2 \\ \dot{x}_3 \end{bmatrix} = A \begin{bmatrix} x_1 \\ x_2 \\ x_3 \end{bmatrix} + B_1 \begin{bmatrix} v_1 \\ v_2 \end{bmatrix} + B_2 \begin{bmatrix} p_1 \\ p_2 \end{bmatrix},$$
$$y = C \begin{bmatrix} x_1 \\ x_2 \\ x_3 \end{bmatrix}, \tag{4.10}$$

where

$$A = \begin{bmatrix} -\frac{f_1}{C_1} & 0 & 0 \\ 0 & -\frac{f_2}{C_2} & 0 \\ \frac{f_1}{C_3} & \frac{f_2}{C_3} & -\frac{f_1+f_2+f_3}{C_3} \end{bmatrix},$$

$$B_1 = \begin{bmatrix} 0 & 0 \\ \frac{1}{C_2} & 0 \\ 0 & \frac{1}{C_3} \end{bmatrix}, B_2 = \begin{bmatrix} 0 & 0 \\ \frac{f_2}{C_2} & 0 \\ 0 & \frac{f_3}{C_3} \end{bmatrix},$$

$$C = \begin{bmatrix} 0 & 0 & 1 \end{bmatrix},$$

$p_1 = t_{02}$, $p_2 = t_{03}$ are the process perturbations, $\nu_1 = w_2$, $\nu_2 = w_3$ are the process inputs, $\chi_1 = t_1, \chi_2 = t_2, \chi_3 = t_3$ are the process states, and y is the process output. If $f_1 = f_2 = f_3 = 1\ \text{m}^3/\text{s}$ and $C_1 = C_2 = C_3 = 1000\ \text{kcal/m}^3\,{}^\circ\text{C}$, so the process matrices are as follows:

$$A = \begin{bmatrix} -\frac{1}{1000} & 0 & 0 \\ 0 & -\frac{1}{1000} & 0 \\ \frac{1}{1000} & \frac{1}{1000} & -\frac{3}{1000} \end{bmatrix},$$

$$B = B_1 = B_2 = \begin{bmatrix} 0 & 0 \\ \frac{1}{1000} & 0 \\ 0 & \frac{1}{1000} \end{bmatrix}, \tag{4.11}$$

$$C = \begin{bmatrix} 0 & 0 & 1 \end{bmatrix}.$$

The controllability of the process (4.10) and (4.11) is analyzed as:

$$\mathbb{R} = \begin{bmatrix} B \\ AB \\ A^2B \end{bmatrix}$$

$$= \begin{bmatrix} 0 & 0 & 0 & 0 & 0 & 0 \\ \frac{1}{1000} & 0 & -\frac{1}{1000000} & 0 & \frac{1}{1000000000} & 0 \\ 0 & \frac{1}{1000} & \frac{1}{1000000} & -\frac{3}{1000000} & -\frac{1}{250000000} & \frac{9}{1000000000} \end{bmatrix}. \tag{4.12}$$

Since the rank of $\mathbb{R}$ in (4.12) is 2, 2 states of the process are able to be regulated and 1 state of the process is notable to be regulated, it will be seen in the experiments.

In this experiment, the mathematical model of the thermal process is given by the equality (4.1) and (4.10), 39.15, 39.15, and 39.15 °C are chosen as the initial conditions for the process states χ_1, χ_2, and χ_3. The perturbations are p_1, p_2 are real data taken from 0 to 10 s. The reference output is $y_r = 39.25$ °C.

The purpose of the GR is that the process output y of (4.1) and (4.10) should converge to the reference output y_r of (4.2) with the regulator of [6] using the output $y = \chi_3$. The GR is given with the factor $L = \begin{bmatrix} 1 & 1 & 1 \end{bmatrix}^T$.

The purpose of the SMR is that the process output y of (4.1) and (4.10) should converge to the reference output y_r of (4.2) with the regulator of (4.4) using the output $y = \chi_3$. The SMR is given by the equalities (4.4) with the factors $B^* = \begin{bmatrix} 0 & 1000 & 0 \\ 0 & 0 & 1000 \end{bmatrix}$, $L = \begin{bmatrix} 1 \\ 1 \\ 1 \end{bmatrix}$, $K = \begin{bmatrix} 0.04 \\ 0.04 \\ 0.04 \end{bmatrix}$. The Lyapunov equality (4.8) is satisfied with $T = \begin{bmatrix} 333.67 & -166.33 & -166.33 \\ -166.33 & 333.67 & -166.33 \\ -166.33 & -166.33 & 332.17 \end{bmatrix}$, $W = \begin{bmatrix} 1.0 & 0.0 & 0.01251 \\ 0.0 & 1.0 & 0.01251 \\ 0.01251 & 0.01251 & 1.013 \end{bmatrix}$. Taking into account Theorem 1, the states of the regulator are asymptotic stable with factors $\rho = 982.32$ and $\sigma = 1.9822 \times 10^{-3}$.

Figures 4.4–4.6 show the perturbations, states, and output for the regulators, where P are the perturbations, DR are the desired references, GR is the geometric regulator, and SMR is the sliding-mode regulator. Table 4.1 shows the MSE for the regulators using (4.9).

From Figure 4.4–4.6, it can be seen that the SMR improves the GR due to the output and states in the first get better perturbation attenuation than in the second. Also, it can be seen that states 2 and 3 are regulated while the state 1 is

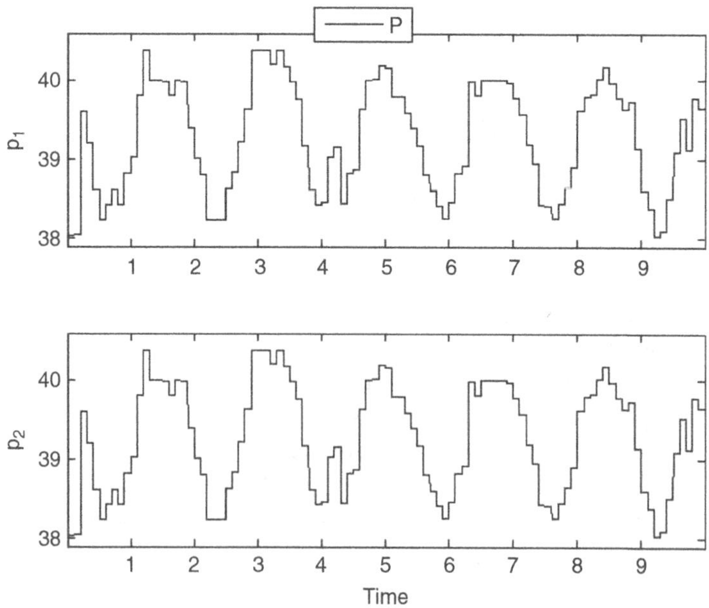

Figure 4.4 Perturbations for the experiment 1.

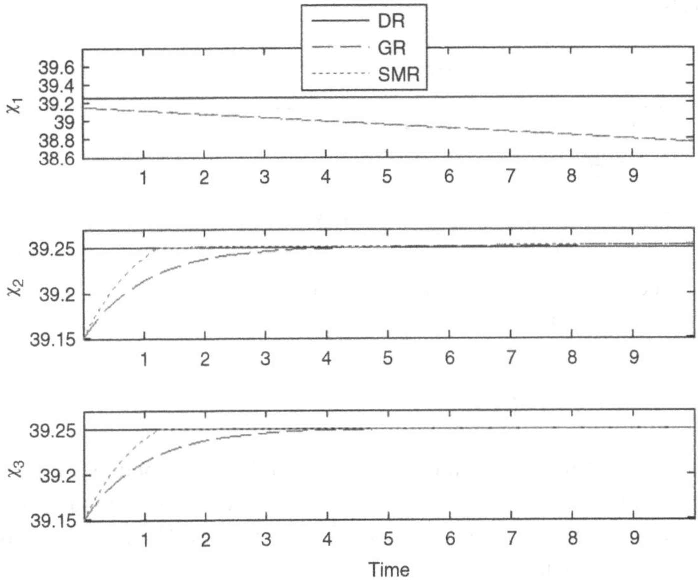

Figure 4.5 States for the experiment 1.

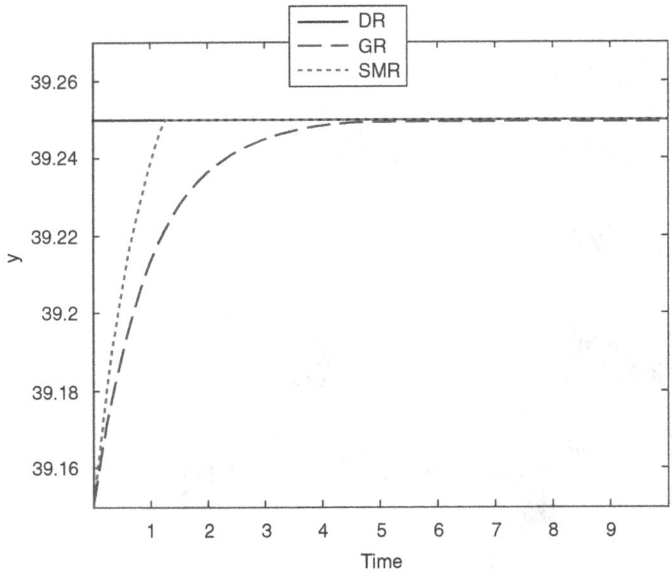

Figure 4.6 The output for the experiment 1.

Table 4.1 The MSE for the regulators.

Strategies	MSE for $\tilde{y}$	MSE for $\tilde{\chi}$
GR	0.0159	0.2244
SMR	0.0123	0.2239

not regulated; it is due to the analysis (4.12), which proved that the process is not completely able to be regulated. From Table 4.1, it can be shown that the SMR achieves better accuracy when compared with the GR due to the MSE for the first is smaller than for the second.

4.5 The Hydraulic Process

This process consists of three interconnected tanks [5]. It is understood that a fixed amount of liquid enters the tank; it sends the tank liquid to a determined liquid level. The valves allow a discharge of liquid outside the process. The purpose is to keep the liquid levels in the tanks at a desired value.

Figure 4.7 shows the liquid-level process, which is employed to get the real data of perturbations and to validate a model. Angular keys with barrel type of 1/2 inch are adapted as valves. Servo motors of 0.1 revolutions per minute are employed to

Figure 4.7 The liquid-level process. *Source*: Rubio et al. [29].

move the valves. Ultrasonic sensors HC-SR04 are employed to get the level measures. An Arduino microcontroller is employed to send the level measures to a personal computer. The personal computer is employed to receive and to save the level measures.

Figure 4.8 shows the liquid-level process diagram, where R_1, R_2, R_3 are the resistances of the valves 1, 2, 3 (m^2/s), C_1, C_2, C_3 are the capacitances of the tanks 1, 2, 3 (m^2), h_1, h_2, h_3 are the liquid levels in the tanks 1, 2, 3, and q_1, q_2 are the liquid inputs to the tanks 1, 2.

The model of the hydraulic process in state space is:

$$\begin{bmatrix} \dot{x}_1 \\ \dot{x}_2 \\ \dot{x}_3 \end{bmatrix} = A \begin{bmatrix} x_1 \\ x_2 \\ x_3 \end{bmatrix} + B \begin{bmatrix} v_1 \\ v_2 \end{bmatrix} + B \begin{bmatrix} p_1 \\ p_2 \end{bmatrix}, \qquad y = C \begin{bmatrix} x_1 \\ x_2 \\ x_3 \end{bmatrix}, \tag{4.13}$$

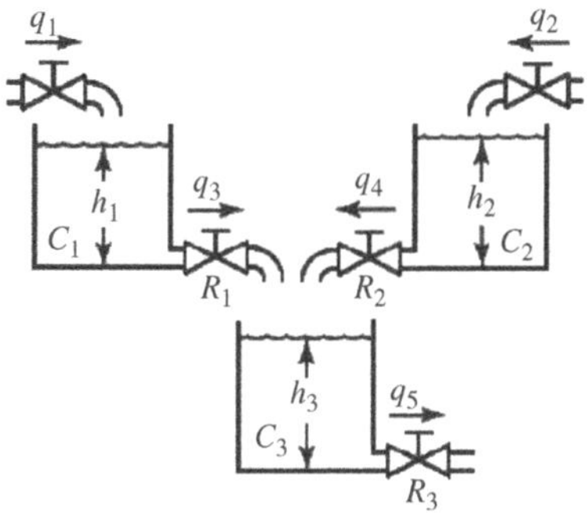

Figure 4.8 The liquid-level process diagram.

where:

$$A = \begin{bmatrix} -\frac{1}{R_1C_1} & 0 & 0 \\ 0 & -\frac{1}{R_2C_2} & 0 \\ \frac{1}{R_1C_3} & \frac{1}{R_2C_3} & -\frac{1}{R_3C_3} \end{bmatrix},$$

$$B = \begin{bmatrix} \frac{1}{C_1} & 0 \\ 0 & \frac{1}{C_2} \\ 0 & 0 \end{bmatrix}, \tag{4.14}$$

$$C = \begin{bmatrix} 0 & 0 & 1 \end{bmatrix},$$

$\chi_1 = h_1$, $\chi_2 = h_2$, $\chi_3 = h_3$ are the process states, $\nu_1 = q_1$, $\nu_2 = q_2$ are the process inputs. If $R_1 = R_2 = R_3 = 0.5\,\text{m}^2/\text{s}$ and $C_1 = C_2 = C_3 = 0.5\,\text{m}^2$; so, the process matrices are:

$$A = \begin{bmatrix} -4 & 0 & 0 \\ 0 & -4 & 0 \\ 4 & 4 & -4 \end{bmatrix},$$

$$B = \begin{bmatrix} 2 & 0 \\ 0 & 2 \\ 0 & 0 \end{bmatrix},$$

$$C = \begin{bmatrix} 0 & 0 & 1 \end{bmatrix},$$

The controllability of the process (4.13) and (4.14) is analyzed as:

$$\mathbb{R} = \begin{bmatrix} B \\ AB \\ A^2B \end{bmatrix}$$

$$= \begin{bmatrix} \frac{1}{0.5} & 0 & -8.0 & 0 & 32.0 & 0 \\ 0 & \frac{1}{0.5} & 0 & -8.0 & 0 & 32.0 \\ 0 & 0 & 8.0 & 8.0 & -64.0 & -64.0 \end{bmatrix}. \tag{4.15}$$

Since the rank of $\mathbb{R}$ in (4.15) is 3, the 3 states of the process are able to be regulated, it will be seen in the experiments.

In this experiment, the mathematical model of the hydraulic process is given by the equality (4.1) and (4.13), 0, 0, and 0 m are chosen as the initial conditions for the process states χ_1, χ_2, and χ_3. The perturbations are p_1, p_2 are real data taken from 0 to 10 s. The reference output is $y_r = 0.08$ m.

The purpose of the GR is that the process output y of (4.1) and (4.13) should converge to the reference output y_r of (4.2) with the regulator of [6] using the output $y = \chi_3$. The GR is given with the factor $L = \begin{bmatrix} 10 & 10 & 10 \end{bmatrix}^T$.

The purpose of the SMR is that the process output y of (4.1) and (4.13) should converge to the reference output y_r of (4.2) with the regulator of (4.4) using the output $y = \chi_3$. The SMR is given by the equality (4.4) with the factors $B^* = \begin{bmatrix} \frac{1}{2} & 0 & 0 \\ 0 & \frac{1}{2} & 0 \end{bmatrix}$, $L = \begin{bmatrix} 10 \\ 10 \\ 10 \end{bmatrix}$, $K = \begin{bmatrix} 0.08 \\ 0.08 \\ 0.08 \end{bmatrix}$. The Lyapunov equality (4.8) is satisfied with $T = \begin{bmatrix} 9.9673 \times 10^{-2} & -2.5327 \times 10^{-2} & -2.5327 \times 10^{-2} \\ -2.5327 \times 10^{-2} & 9.9673 \times 10^{-2} & -2.5327 \times 10^{-2} \\ -2.5327 \times 10^{-2} & -2.5327 \times 10^{-2} & 7.1895 \times 10^{-2} \end{bmatrix}$, $W = \begin{bmatrix} 1.0 & 0.0 & -6.0 \times 10^{-6} \\ 0.0 & 1.0 & -6.0 \times 10^{-6} \\ -6.0 \times 10^{-6} & -6.0 \times 10^{-6} & 0.99998 \end{bmatrix}$. Taking into account Theorem 1, the states of the regulator are asymptotic stable with factors $\rho = 3.3528$ and $\sigma = 7.9997$.

Figures 4.9–4.11 show the perturbations, states, and output for the regulators where P are the perturbations, DR are the desired references, GR is the geometric regulator, and SMR is the sliding-mode regulator. Table 4.2 shows the MSE for the regulators using (4.9).

From Figures 4.9 to 4.11, it can be seen that the SMR improves the GR due to the output and states in the first get better perturbation attenuation than in the second. Also, it can be seen that states 1, 2, and 3 are regulated; it is due to the analysis (4.15), which proved that the process is completely able to be regulated. From Table 4.2, it can be shown that the SMR achieves better accuracy when compared with the GR due to the MSE for the first is smaller than for the second.

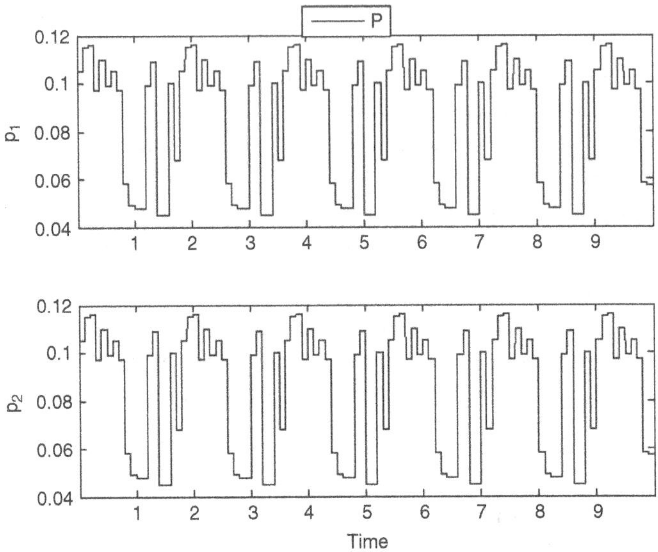

Figure 4.9 Perturbations for the experiment 2.

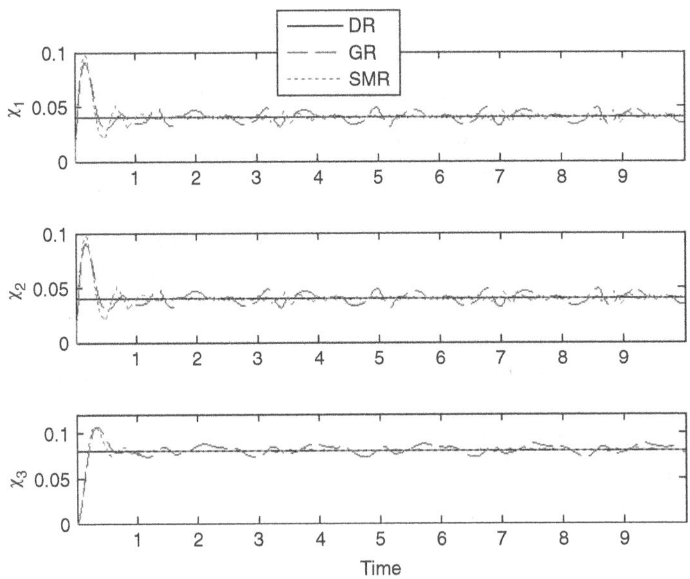

Figure 4.10 States for the experiment 2.

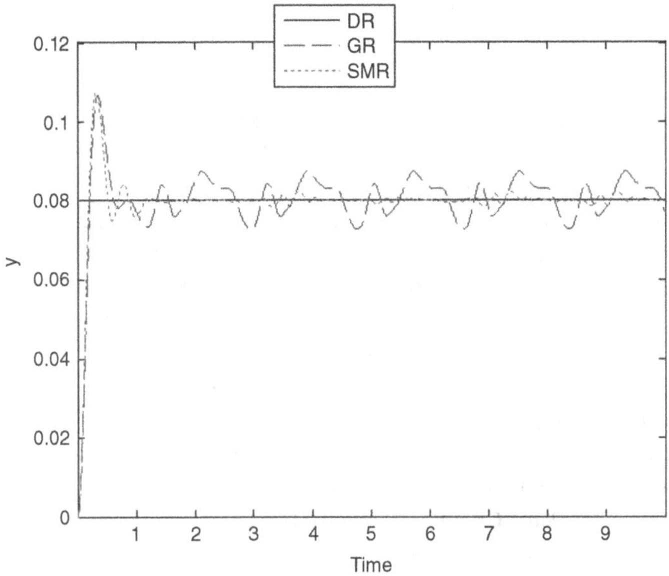

Figure 4.11 The output for the experiment 2.

Table 4.2 The MSE for the regulators.

Strategies	MSE for $\tilde{y}$	MSE for $\tilde{\chi}$
GR	0.0066	0.0104
SMR	0.0058	0.0097

4.6 Concluding Remarks

In this chapter, the usefulness of the sliding-mode approach is recommended for solving the perturbation attenuation issue. The comparisons showed that the SMR achieves better perturbation attenuation when is compared with the geometric regulator. It is important to remark that the perturbations measurement is not required. This chapter represents the theoretical and application basis for developing the perturbation attenuation issues, for being able afterward to perform an implementation in functioning processes. In the future, other kinds of regulators will be designed for the perturbation attenuation, the adaptive learning will be combined with the recommended regulator, or the trajectory tracking issue will be studied.

References

1 Leite D. F., Hell M. B., Costa Jr. P., Gomide F. (2009), Real-time fault diagnosis of nonlinear systems, Nonlinear Analysis, 71 (12), e2665–e2673.
2 Lemos A., Caminhas W., Gomide F., (2013), Adaptive fault detection and diagnosis using an evolving fuzzy classifier, Information Sciences, 220, 64–85.
3 Wu H.-T., Jiao C.-Q., Cui X., Liu X.-F., Ji J.-F. (2017), Transient electromagnetic disturbance induced on the ports of intelligent component of electronic instrument transformer due to switching operations in 500 kV GIS substations, IEEE Access, 5, 5104–5112.
4 Yin S., Gao H., Qiu J., Kaynak O. (2017), Fault detection for nonlinear process with deterministic disturbances: a just-in-time learning based data driven method, IEEE Transactions on Cybernetics, 47 (11), 3649–3657.
5 Rubio J. J., Melendez F., Figueroa M. (2014), An observer with controller to detect and reject disturbances, International Journal of Control, 87 (3), 524–536.
6 Rubio J. J., Figueroa M., Perez-Cruz J. H., Bejarano F. J. (2012), Geometric approach and structure at infinite controls for the disturbance rejection, IET Control Theory & Applications, 6 (16), 2528–2537.
7 Aguilar-Ibañez C., Sira-Ramirez H., Acosta J. A. (2017), Stability of active disturbance rejection control for uncertain systems: a Lyapunov perspective, International Journal of Robust and Nonlinear Control, 27 (18), 4541–4553.
8 Aguilar-Ibañez C., Sira-Ramirez H., Suarez-Castanon M. S. (2016), A linear active disturbance rejection control for a ball and rigid triangle system, Mathematical Problems in Engineering, 2016, 1–11.
9 Li T., Liu A., Huang C. (2016), A similarity scenario-based recommendation model with small disturbances for unknown items in social networks, IEEE Access, 4, 9251–9272.
10 Sun T., Zhang J., Pan Y. (2017), Active disturbance rejection control of surface vessels using composite error updated extended state observer, Asian Journal of Control, 19 (6), 1–10.
11 Han H., Chen J., Reza Karimi H. (2017), State and disturbance observers-based polynomial fuzzy controller, Information Sciences, 382-383, 38–59.
12 Selvaraj P., Kaviarasan B., Sakthivel R., Reza Karimi H., (2017), Fault-tolerant SMC for Takagi–Sugeno fuzzy systems with time-varying delay and actuator saturation, IET Control Theory & Applications, 11 (8), 1112–1123.
13 Selvaraj P., Sakthivel R., Reza Karimi H. (2016), Equivalent-input-disturbance-based repetitive tracking control for Takagi–Sugeno fuzzy systems with saturating actuator, IET Control Theory & Applications, 10 (15), 1916-1927.
14 Diao X., Jin Y., Ma L., Ding S., Jiang H. (2017), Composite active front steering controller design for vehicle system, IEEE Access, 5, 6697–6706.
15 Liu M., Zhang L., Shi P., Reza Karimi H. (2014), Robust control of stochastic systems against bounded disturbances with application to flight control, IEEE Transactions on Industrial Electronics, 61 (3), 1504–1515.

16 Reza Karimi H., Niu Y., Rossell J. M., Yang F. (2017), Analysis and synthesis of control systems over wireless digital channels, Journal of the Franklin Institute, 354, 3649–3653.

17 Zhang J.-E. (2017), Robustness analysis of global exponential stability of nonlinear systems with deviating argument and stochastic disturbance, IEEE Access, 5, 13446–13454.

18 Pan Y., Er M. J., Huang D., Sun T. (2012), Practical adaptive fuzzy H∞ tracking control of uncertain nonlinear systems, International Journal of Fuzzy Systems, 14 (4), 463–473.

19 Pan Y., Zhou Y., Sun T., Er M. J. (2013), Composite adaptive fuzzy H∞ tracking control of uncertain nonlinear systems, Neurocomputing, 99, 15–24.

20 Pan M., Zhang K., Chen Y.-H., Huang J. (2017), A new robust tracking control design for turbofan engines: H∞/Leitmann approach, Applied Sciences, 7 (439), 1–21.

21 Wei Y., Qiu J., Reza Karimi H., Wang M. (2014), New results on H∞ dynamic output feedback control for Markovian jump systems with time-varying delay and defective mode information, Optimal Control Applications and Methods, 35, 656–675.

22 Wei Y., Qiu J., Reza Karimi H. (2015), Quantized H∞ filtering for continuous-time Markovian jump systems with deficient mode information, Asian Journal of Control, 17 (5), 1914–1923.

23 Gao Y., Li H., Wu L., Reza Karimi H., Lam H.-K. (2016), Optimal control of discrete-time interval type-2 fuzzy-model-based systems with d-stability constraint and control saturation, Signal Processing, 120, 409–421.

24 Liu D., Wu C., Zhou Q., Lam H.-K. (2015), Fuzzy guaranteed cost output tracking control for fuzzy discrete-time systems with different premise variables, Wiley Periodicals, 21 (5), 265–276.

25 Pan Y., Yu H., (2015), Dynamic surface control via singular perturbation analysis, Automatica, 57, 29–33.

26 Wang X., Zhao H., Sun Q., Chen Y. H. (2017), A new high-order adaptive robust control for constraint following of mechanical systems, Asian Journal of Control, 19 (5), 1–16.

27 Aguilar-Ibañez C. (2017), Stabilization of the PVTOL aircraft based on a sliding mode and a saturation function, International Journal of Robust and Nonlinear Control, 27, 843–859.

28 Wang J., Gao Y., Qiu J., Ahn C. K. (2017), Sliding mode control for non-linear systems by Takagi-Sugeno fuzzy model and delta operator approaches, IET Control Theory & Applications, 11 (8), 1205–1213.

29 Rubio J. J., Soriano E., Juarez C. F., Pacheco J. (2017), Sliding mode regulator for the perturbations attenuation in two tank plants, IEEE Access, 5 (1), 20504–20511.

5

Uniform Stable Observer for the Perturbation Estimation in Two Energy Processes

5.1 Introduction

The perturbations are undesired signals presented as inputs of a process which affect the outputs. This problem has been studied in many processes. Estimation of perturbations in the process output is required since they can affect the sensors, actuators, plants, or regulators, causing accidents or unnecessary costs. So, an observer to estimate the process perturbations is an important and current issue in the theory and applications.

There is some research about the perturbation attenuation regulator. In [1], an active perturbation attenuation methodology applied to a large class of uncertain flat processes is mentioned. A problem on bounded functional state estimation for time-delay processes with unknown bounded perturbations is recommended in [2, 3]. In [4, 5], the geometric regulator for the perturbation attenuation is mentioned. The methodology of the active perturbation attenuation regulator and its theoretical analysis are reviewed in [6]. In [7], the structure at infinity of infinite-dimensional linear time-invariant processes with input and output spaces is discussed. In [8], local regulators are designed such that the process achieves the output synchronization with a desired perturbation attenuation performance. Regulator of a platoon of vehicles is designed in [9] when some suffer from perturbations. In [10], the perturbation estimations are considered into a virtual regulator law in each step to compensate the mismatched perturbations. The geometric approach is highly explained in [11]. In [12], the author reviews the status of geometric state space theory as developed for application of linear processes. A novel solution for an electro-hydraulic variable valve timing process based on the concept of active perturbation attenuation regulator is proposed in [13]. In [14], the authors study a singular perturbation margin and a generalized gain margin for linear time-invariant processes from the view of perturbations.

The problem of a composite hierarchical anti-perturbations regulator is addressed in [15]. In [16], the problems of the composite perturbations–observer-based output feedback regulator and passive regulator for the Markovian jump processes with multiple perturbations are considered. A novel modification of active perturbation attenuation regulator is proposed in [17] so that good perturbation attenuation is achieved. An active perturbation attenuation regulator has been designed in [18] for plants with delay. In [19], the problem of aperiodic-perturbation estimation and attenuation in a modified repetitive-regulator process is analyzed.

From the aforementioned research, in [1, 3, 6, 10, 13, 15–19], the authors have used observers for the perturbation estimation; so, it should be interesting to analyze the stability and convergence of this observer for the estimation of the states and perturbations.

In this chapter, an alternative observer is recommended for the state and perturbation estimation in disturbed processes. The benefit is to present some methods to find the gains of the proposed observer to warrant its stability and the convergence of its errors with the final benefit of obtaining a good estimation. The recommended strategy is applied for the state and perturbation estimation in a wind turbine and in an electric vehicle. The wind turbine has a rotatory tower and the electric vehicle has generators connected with the wheels. The first is for attracting the incoming air to be become in electricity, and the other for attracting the vehicle movement to be become in electricity.

The rest of the chapter is organized as follows. Section 5.2 shows the observer and its stability analysis for the processes without perturbations. Section 5.3 presents the observer, its stability, convergence analysis, and perturbation estimation for the processes with perturbations. Sections 5.4 and 5.5 use the recommended strategy to estimate the states and perturbations in a wind turbine and an electric vehicle, respectively. Section 5.6 presents the conclusion and future research.

5.2 Observer for the Processes Without Perturbations

Consider the following not disturbed process (A, B, C, D):

$$\begin{aligned} \dot{\chi} &= A\chi + B\nu, \\ y &= C\chi + D\nu, \end{aligned} \tag{5.1}$$

where $t \geq 0$, $\chi \in \Re^n$ is the state, $\nu \in \Re^m$ is the regulator input, $y \in \Re^p$ is the output. $A \in \Re^{n\times n}$, $B \in \Re^{n\times m}$, $C \in \Re^{p\times n}$, and $D \in \Re^{p\times m}$ are matrices representing the maps $\mathbf{A} : \Re^n \to \Re^n$, $\mathbf{B} : \Re^m \to \Re^n$, $\mathbf{C} : \Re^n \to \Re^p$, and $\mathbf{D} : \Re^m \to \Re^p$, respectively.

5.2.1 Observer Design

A device or even a program that estimates the state variables of a process is named observer. If the observer estimates all the n state variables, it is named full-order observer.

The mathematical model of the observer, that is, the dynamics of the estimated state, is as follows:

$$\begin{aligned} \dot{\hat{\chi}} &= A\hat{\chi} + Bv + L(\tilde{y}), \\ \hat{y} &= C\hat{\chi} + Dv, \end{aligned} \tag{5.2}$$

where $\hat{\chi}$ is the estimated state vector, $\hat{y}$ is the estimated output and the term L is a function named $L : \mathfrak{R}^p \to \mathfrak{R}^n$, whose matrix representation is named output-injection matrix. $\tilde{y} = y - \hat{y}$ is the output error.

The observer error is the difference between the real state and the state estimated by the observer, it is $\tilde{\chi} = \chi - \hat{\chi}$.

5.2.2 Stability Analysis of the Observer

In this section, the error of the observer applied to the not disturbed process is warranted to be exponentially stable based in the solution of the Lyapunov method.

From (5.1) and (5.2), the following equality is obtained:

$$\tilde{y} = C\tilde{\chi}. \tag{5.3}$$

From (5.1) and (5.2), and the observer error equation can be formed as follows:

$$\dot{\tilde{\chi}} = A\tilde{\chi} - L\tilde{y} = [A - LC]\,\tilde{\chi} = A_0\tilde{\chi}, \tag{5.4}$$

where $A_0 = A - LC$.

Theorem 5.1 shows the stability of the observer.

Theorem 5.1: *The error of the observer (5.8) and 5.9) used to estimate the parameters χ of the process (5.1) is exponentially stable; so, the observer error $\tilde{\chi}$ satisfies:*

$$\|\tilde{\chi}\|^2 \leq \gamma e^{-\beta t}\|\tilde{\chi}_0\|^2, \tag{5.5}$$

where $\tilde{\chi}_0$ is the initial condition of $\tilde{\chi}$, $\gamma = \frac{\lambda_{\max}(P)}{\lambda_{\min}(P)}$, $\beta = \lambda_{\min}(QP^{-1})$, $\|\cdot\|$ is the Euclidean norm in $\mathfrak{R}^n$, $P \in \mathfrak{R}^{n\times n}$ and $Q \in \mathfrak{R}^{n\times n}$ are positive matrices which satisfy the following equation:

$$A_0^T P + PA_0 = -Q, \tag{5.6}$$

where A_0 is given in (5.4).

Proof. See [20] for the proof. Section 5.3 shows the observer for the processes with perturbations. ■

5.3 Observer for the Processes with Perturbations

Consider the following disturbed process (A, B, C, D, E):

$$\begin{aligned} \dot{\chi} &= A\chi + f(\chi, \nu) + Ed \\ y &= C\chi, \end{aligned} \tag{5.7}$$

where $t \geq 0$, $\chi \in \Re^n$ is the state, $\nu \in \Re^m$ is the regulator input, $y \in \Re^p$ is the output, $d \in \Re^m$ are the perturbations, $f(\chi, \nu)$ is a nonlinear map. $A \in \Re^{n\times n}$, $B \in \Re^{n\times m}$, $C \in \Re^{p\times n}$, $D \in \Re^{p\times m}$, and $E \in \Re^{n\times m}$ are matrices representing the maps $\mathbf{A} : \Re^n \to \Re^n$, $\mathbf{B} : \Re^m \to \Re^n$, $\mathbf{C} : \Re^n \to \Re^p$, and $\mathbf{E} : \Re^m \to \Re^n$, respectively.

5.3.1 Observer Design

In this section, an observer will be designed based on the assumption that the input ν and output y are known, but the perturbations d is unknown. Let $\hat{\chi} \in \Re^n$ be the estimation of the states $\chi \in \Re^n$. Define the output error $\tilde{y}$ as follows:

$$\tilde{y} = y - \hat{y} = C\tilde{\chi}, \tag{5.8}$$

where $y = C\chi + D\nu \in \Re^p$, $\hat{y} = C\hat{\chi} \in \Re^p$, $\tilde{\chi} = \chi - \hat{\chi} \in \Re^n$ is the observer error.

The following observer is proposed:

$$\begin{aligned} \dot{\hat{\chi}} &= A\hat{\chi} + f(\hat{\chi}, \nu) + L\tilde{y}, \\ \hat{y} &= C\hat{\chi}, \end{aligned} \tag{5.9}$$

where the parameters, L is a vector or matrix such that $LC \in \Re^{n\times n}$ is constant.

The benefit of the observer is that the observer states $\hat{\chi}$ and perturbations $\hat{d}$ of (5.9) may follow the plant states χ and perturbations d of (5.7) using the output y in the output error (5.8). See Figure 5.1.

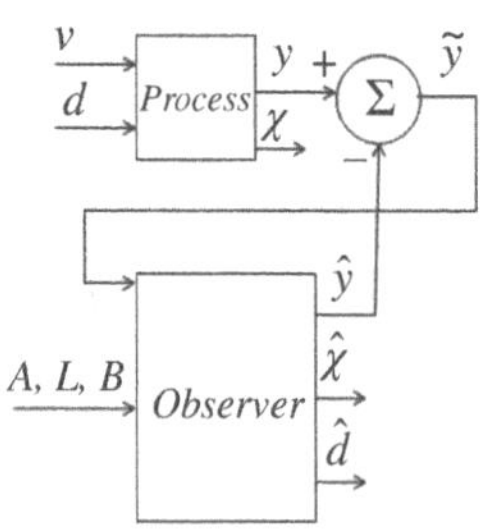

Figure 5.1 Perturbations observer.

Remark 5.1: Note that the perturbations d are unknown in the proposed observer; it only requires knowledge of the input ν and output y.

5.3.2 Stability Analysis of the Observer

In this section, the error of the observer applied to the disturbed process is warranted to be uniform and stable based in the solution of the Lyapunov method.

From (5.7–5.9), the observer error equation can be formed as follows:

$$\begin{aligned}\dot{\widetilde{\chi}} &= A\widetilde{\chi} + \widetilde{f} + Ed - L\widetilde{y} \\ &= [A - LC]\,\widetilde{\chi} + \widetilde{f} + Ed \\ &= A_0\widetilde{\chi} + \widetilde{f} + Ed,\end{aligned} \tag{5.10}$$

where $\widetilde{f} = f(\widehat{\chi}, \nu) - f(\chi, \nu)$, $A_0 = A - LC$.

Theorem 5.2 shows the stability of the proposed observer.

Theorem 5.2: *The error of the observer (5.8) and (5.9) used to estimate the states χ and perturbations d of the disturbed process (5.7) is uniform stable; so, the observer error $\widetilde{\chi}$ satisfies:*

$$\|\widetilde{\chi}\|^2 \le \gamma e^{-\beta t}\|\widetilde{\chi}_0\|^2 + \xi, \tag{5.11}$$

where $\widetilde{\chi}_0$ is the initial condition of $\widetilde{\chi}$, $\gamma = \frac{\lambda_{\max}(P)}{\lambda_{\min}(P)}$, $\xi = \frac{\alpha}{\beta\lambda_{\min}(P)}$, $\frac{2}{\theta}E^T P E\overline{d}^2 + \frac{2}{\theta}\overline{f}^T P\overline{f}$, $\beta = \lambda_{\min}(QP^{-1})$, $|d^2| \le \overline{d}^2$, $|\widetilde{f}| \le \overline{f}$, $\|\cdot\|$ is the Euclidean norm in $\Re^n$ and $|\cdot|$ is the absolute value, $0 < \gamma \in \Re$, $0 < \beta \in \Re$, $0 < \xi \in \Re$, $P \in \Re^{n\times n}$ and $Q \in \Re^{n\times n}$ are positive matrices which satisfy the following equation:

$$A_0^T P + PA_0 + \theta P = -Q, \tag{5.12}$$

where A_0 is given in (5.10), $0 < \theta \in \Re$ is selected by the designer.

Proof. See [20] for the proof. ■

Remark 5.2: If the parameter α can be arbitrarily small, the uniform stability is warranted because of its definition. From (5.11), the parameter $\alpha = \frac{1}{\theta}E^T P E\overline{d}^2$ can become arbitrarily small by increasing the value of the selected parameter θ, or by decreasing the values of the selected parameter P.

Remark 5.3: Note that θ should be selected small such that (5.12) is solved, but it also should be selected big for giving α to warrant the uniform stability. Please also note that θ is not required in the observer (5.9).

5.3.3 Convergence Analysis of the Observer

Theorem 5.3 presents the convergence of the state estimation in the recommended observer.

Theorem 5.3: *If the real parts of the eigenvalues in A_0 are negative, so the error $\widetilde{\chi}$ of the observer (5.8) and (5.9) used to estimate the states and perturbations of the disturbed process (5.7) will converge to:*

$$\lim_{t\to\infty} |\widetilde{\chi}| \leq \overline{\chi} \tag{5.13}$$

where $\overline{\chi} = \left|-A_0^{-1}\left(\overline{f} + E\overline{d}\right)\right|$, A_0^{-1} is the inverse of A_0, A_0 is given in (5.10), E and d are given in (5.7), $|d| \leq \overline{d}$, $\left|\widetilde{f}\right| \leq \overline{f}$, $|\cdot|$ is the absolute value.

Proof. See [20] for the proof. ■

5.3.4 Estimation of the Perturbations

From (5.13), the following fact is true:

$$\lim_{t\to\infty} |\widetilde{\chi}| \cong \overline{\chi}, \tag{5.14}$$

where $\overline{\chi} = \left|-A_0^{-1}\left(\overline{f} + E\overline{d}\right)\right|$ is constant. Consequently, $\lim_{t\to\infty} \dot{\widetilde{\chi}} \cong 0$. Substituting in the first equation of (5.10) gives:

$$\begin{aligned}
&\lim_{t\to\infty} \dot{\widetilde{\chi}} = \lim_{t\to\infty} A\widetilde{\chi} + \lim_{t\to\infty} Ed + \lim_{t\to\infty} \widetilde{f} - \lim_{t\to\infty} L\widetilde{y} \\
&\Longrightarrow 0 = \lim_{t\to\infty} A \lim_{t\to\infty} \widetilde{\chi} + \lim_{t\to\infty} Ed + \lim_{t\to\infty} \widetilde{f} - \lim_{t\to\infty} L\widetilde{y} \\
&\Longrightarrow 0 \cong \lim_{t\to\infty} A\overline{\chi} + \lim_{t\to\infty} Ed + \lim_{t\to\infty} \overline{f} - \lim_{t\to\infty} L\widetilde{y} \\
&\Longrightarrow \lim_{t\to\infty} d \cong \lim_{t\to\infty} E^{-1}\left[L\widetilde{y} - A\overline{\chi} - \overline{f}\right],
\end{aligned} \tag{5.15}$$

where E^{-1} is the pseudo-inverse of E. As all the elements of (5.15) are bounded independently of the value of t, it can be rewritten as follows:

$$\widehat{d} = E^{-1}\left[L\widetilde{y} - A\overline{\chi} - \overline{f}\right], \tag{5.16}$$

where $\widehat{d}$ are the estimation of the perturbations d.

Remark 5.4: Note that the disturbed process (5.7) is not required to be stable to obtain satisfactory results of this method.

In the following two sections, the root mean square error (RMSE) will be used for the comparisons, it is given as:

$$J = \left(\frac{1}{T} \int_0^T e^2 d\tau \right)^{\frac{1}{2}}, \tag{5.17}$$

where $J = J_\chi$ and $e^2 = \widetilde{\chi}^2$ for the states, $J = J_y$ and $e^2 = \widetilde{y}^2$ for the output, or $J = J_d$ and $e^2 = \widetilde{d}^2 = \left(d - \widehat{d}\right)^2$ for the perturbations.

5.4 Wind Turbine

Figure 5.2 shows the wind turbine. The wind turbine is employed to catch the incoming air to transform it in electricity. A windward wind turbine of three blades with a rotatory tower is considered. This model is divided into four parts: the first is the mechanic model, the second is the aerodynamic model, the third is the electric model, and finally, the fourth is the combination of the aforementioned models. Some important regulators of turbines are designed in [21–23].

The state variables are $\chi_1 = i_2$, $\chi_2 = \theta_2$, $\chi_3 = \dot{\theta}_2$, $\chi_4 = i_1$, $\chi_5 = \theta_1$, $\chi_6 = \dot{\theta}_1$, the inputs are $\nu_1 = F_2, \nu_2 = V_1$, and the output is $y = V_2 + V_1 = R_e i_2 + V_1$. The analytic model is given as follows [24]:

$$\dot{\chi}_1 = -\frac{(R_2 + R_e)}{L_2}\chi_1 + \frac{k_2}{L_2}\chi_3,$$

$$\dot{\chi}_2 = \chi_3,$$

$$\dot{\chi}_3 = -\frac{k_{b2}}{\left[J_2 + m_2 l_{c2}^2\right]}\chi_2 - \frac{b_{b2}}{\left[J_2 + m_2 l_{c2}^2\right]}\chi_3 - \frac{2\pi k_{b2}}{3\left[J_2 + m_2 l_{c2}^2\right]} + \frac{\cos(\chi_5)}{3\left[J_2 + m_2 l_{c2}^2\right]}\nu_1,$$

$$\dot{\chi}_4 = -\frac{R_1}{L_1}\chi_4 - \frac{k_1}{L_1}\chi_6 + \frac{1}{L_1}\nu_2,$$

$$\dot{\chi}_5 = \chi_6,$$

Figure 5.2 Wind turbine.

$$\dot{\chi}_6 = -\frac{k_{b1}}{\left[J_1 + 1.5m_2l_{c2}^2\right]}\chi_5 - \frac{b_{b1}}{\left[J_1 + 1.5m_2l_{c2}^2\right]}\chi_6 + \frac{k_m}{3\left[J_1 + 1.5m_2l_{c2}^2\right]}\chi_4,$$

$$y = R_e\chi_1 + \nu_2,$$

$$\nu_1 = F_{2a} + F_{2b} + F_{2c},$$

$$F_{2a} = \frac{1}{2\chi_3}\rho A C_p(\lambda, \beta)V_\omega^3,$$

$$F_{2b} = \frac{1}{2\chi_3}\rho A C_p(\lambda, \beta)V_\omega^3,$$

$$F_{2c} = \frac{1}{2\chi_3}\rho A C_p(\lambda, \beta)V_\omega^3,$$

$$\lambda = \frac{\chi_3 R}{V_\omega}, \tag{5.18}$$

where $C_p(\lambda, \beta) = c_1 \left(\frac{c_2}{\lambda_i} - c_3\beta - c_4 \right) e^{-c_5/\lambda_i} + c_6\lambda$, $\frac{1}{\lambda_i} = \frac{1}{\lambda + 0.08\beta} - \frac{0.035}{\beta^3 + 1}$. $c_1 = 0.5176$, $c_2 = 116$, $c_3 = 0.4$, $c_4 = 5$, $c_5 = 21$, and $c_6 = 0.0068$ are coefficients. Table 5.1 shows the parameter descriptions for the wind turbine where the motor is used to move the tower, and the generator is moved by the blade. Equation (5.18) describes the assumption that the air goes in one direction, if $\theta_1 = 0$, so the maximum air intake moves the blades of the wind turbine, but if the tower turns to the left or to the right and θ_1 changes, so the wind turbine turns, and the air intake decreases. Table 5.2 shows the parameter values for the wind turbine.

Table 5.1 Parameter descriptions for the wind turbine.

Parameter	Description
θ_1	Angular position of the tower motor in rad
θ_2	Angular position of a blade in rad
l_{c2}	Length of the blade center in m
m_1	Tower mass in kg
m_2	Blade mass in kg
g	Acceleration gravity in m/s^2
l_1	Constant length of the tower in m
l_{c1}	Length of the tower center in m
τ_{1a}, τ_{2a}	Torque of the motor and generator in kg m^2 rad/s^2
k_{b1}, k_{b2}	Spring effect presented when the blade is near to stop in kg m^2/s^2
b_{b1}, b_{b2}	Shock absorber in kg m^2 rad/s
F_{2a}, F_{2b}, F_{2c}	Forces of the air received by the three blades
k_m	Motor magnetic flux constant of the tower in Wb
i_1, i_2	Motor and generator armature current in A
ρ	Air density in kg/m^3
$A = \pi R^2$	Area swept by the rotor blades in m^2
R	Radius of the blade in m
V_ω	Wind speed in m/s
$C_p(\lambda, \beta)$	Performance coefficient of the wind turbine
λ	Tip speed ratio
β	Blade pitch angle in rad
k_1, k_2	Motor and generator back emf constant in V s/rad
R_1, R_2	Motor and generator armature resistance in Ω
L_1, L_2	Motor and generator armature inductance in H

(Continued)

Table 5.1 (Continued)

Parameter	Description
V_1, V_2	Motor and generator armature voltage in V
R_e	Resistance of an external load in Ω
J_1, J_2	Motor and generator inertia in kg m^2

Table 5.2 Parameter values for the wind turbine.

Parameter	Value	Parameter	Value
l_{c2}	0.5 m	R_e	30 Ω
m_2	0.5 kg	k_m	0.09 Wb
k_{b2}	1×10^{-3} kg m^2/s^2	k_{b1}	1×10^{-3} kg m^2/s^2
b_{b2}	1×10^{-1} kg m^{2}rad/s	b_{b1}	1×10^{-1} kg m^{2}rad/s
k_2	0.45 V s/rad	k_1	0.0045 V s/rad
R_2	6.96 Ω	R_1	18 Ω
L_2	6.031×10^{-1} H	L_1	6.031×10^{-1} H
R	l_{c2} m	V_ω	5 m/s
ρ	1.225 kg/m^3	β	0.5 rad
g	9.81 m/s^2	$J_1 = J_2$	0.01273 kg m^2

A variable change is utilized as a transformation to rewrite the nonlinear process of the form (5.18) as a disturbed process of the form (5.7). Define the input as $\nu = \nu_2$, the perturbations as $d = \cos(\chi_5)\nu_1 - 2\pi k_{b2}$; the states and output remain equal; therefore, the following disturbed process is obtained:

$$
\begin{aligned}
\dot{\chi}_1 &= -\frac{(R_2 + R_e)}{L_2}\chi_1 + \frac{k_2}{L_2}\chi_3,\\
\dot{\chi}_2 &= \chi_3,\\
\dot{\chi}_3 &= -\frac{k_{b2}}{\left[J_2 + m_2 l_{c2}^2\right]}\chi_2 - \frac{b_{b2}}{\left[J_2 + m_2 l_{c2}^2\right]}\chi_3 + \frac{1}{3\left[J_2 + m_2 l_{c2}^2\right]}d,\\
\dot{\chi}_4 &= -\frac{R_1}{L_1}\chi_4 - \frac{k_1}{L_1}\chi_6 + \frac{1}{L_1}\nu,\\
\dot{\chi}_5 &= \chi_6,\\
\dot{\chi}_6 &= -\frac{k_{b1}}{\left[J_1 + 1.5 m_2 l_{c2}^2\right]}\chi_5 - \frac{b_{b1}}{\left[J_1 + 1.5 m_2 l_{c2}^2\right]}\chi_6 + \frac{k_m}{3\left[J_1 + 1.5 m_2 l_{c2}^2\right]}\chi_4,\\
y &= R_e\chi_1 + \nu,
\end{aligned}
\tag{5.19}
$$

If the parameter values of Table 5.2 are considered, so the matrices of the process (A, B, C, D) are:

$$A = \begin{bmatrix} -51.335 & 0 & 0.74614 & 0 & 0 & 0 \\ 0 & 0 & 1 & 0 & 0 & 0 \\ 0 & -7.2606\times 10^{-3} & -0.72606 & 0 & 0 & 0 \\ 0 & 0 & 0 & -29.846 & 0 & -7.4614\times 10^{-3} \\ 0 & 0 & 0 & 0 & 0 & 1 \\ 0 & 0 & 0 & 0.14983 & -4.9943\times 10^{-3} & -0.49943 \end{bmatrix}, \quad (5.20)$$

$$B = \begin{bmatrix} 0 & 0 & 0 & 1.6581 & 0 & 0 \end{bmatrix}^T, \quad (5.21)$$

$$C = \begin{bmatrix} 30 & 0 & 0 & 0 & 0 & 0 \end{bmatrix}, \quad (5.22)$$

$$D = 1, \quad (5.23)$$

and the matrix of the perturbations is:

$$E = \begin{bmatrix} 0 & 0 & 2.4202 & 0 & 0 & 0 \end{bmatrix}^T. \quad (5.24)$$

5.4.1 Comparisons

In this section, the observer of this chapter named Observer will be compared with the observer of [25] named Observer2. The mathematical model of the wind turbine with a rotatory tower is given by (5.18). The initial conditions for the plant are $\chi_0 = [1\times 10^{-5}, 1\times 10^{-5}, 1\times 10^{-5}, 1\times 10^{-5}, 1\times 10^{-5}]^T$; therefore, the wind turbine starts near to the rest position. The benefit of the observer is that the estimated states $\hat{\chi}$ must follow the states χ and the estimated perturbations $\hat{d}$ must follow the perturbations d. The perturbations are the incoming air which is defined as follows $d = \cos(\chi_5)\nu_1 - 2\pi k_{b2}$, $\nu_1 = 0.1\sin(2t)$ from 0 s to 20 s. The input of the tower ν_2 is given as $\nu_2 = 3.5\sin(0.315t)$V from 0 s to 20 s. t is the time in s.

Observer2 of [25] is used with initial conditions $\hat{\chi}_i = [4.5\times 10^{-4}, 4.5\times 10^{-4}, 4.5\times 10^{-4}, 4.5\times 10^{-4}, 4.5\times 10^{-4}, 4.5\times 10^{-4}]^T$ and with gains $L_i = [0.16581, 1.0, 24.202, 0.16581, 0.1, 0.16648]^T$.

Observer of this study is employed with initial conditions $\hat{\chi}_0 = [4.5\times 10^{-4}, 4.5\times 10^{-4}, 4.5\times 10^{-4}, 4.5\times 10^{-4}, 4.5\times 10^{-4}, 4.5\times 10^{-4}]^T$ and with gains

$L = [0.16581,\ 2.5,\ 60.505,\ 0.16581,\ 0.001,\ 1.6648 \times 10^{-3}]^T$. The condition of Theorem 5.2 is satisfied with $P = \begin{bmatrix} p_{11} & p_{12} & p_{13} & 0 & 0 & 0 \\ p_{12} & p_{22} & 0 & 0 & 0 & 0 \\ p_{13} & 0 & p_{33} & 0 & 0 & 0 \\ 0 & 0 & 0 & p_{44} & 0 & p_{46} \\ 0 & 0 & 0 & 0 & p_{55} & p_{56} \\ 0 & 0 & 0 & p_{46} & p_{56} & p_{66} \end{bmatrix}$, $p_{11} = 20.76$, $p_{12} = -3.2387 \times 10^{-2}$, $p_{13} = -0.64237$, $p_{22} = 2.4374 \times 10^{-2}$, $p_{33} = 2.8698 \times 10^{-2}$, $p_{44} = 1.6781 \times 10^{-2}$, $p_{46} = 4.9920 \times 10^{-3}$, $p_{56} = 1.6640 \times 10^{-4}$, $p_{55} = 5.1333 \times 10^{-3}$, $p_{66} = 1.0115$, and $Q = \begin{bmatrix} q_{11} & q_{12} & q_{13} & q_{14} & q_{15} & q_{16} \\ q_{12} & q_{22} & q_{23} & 0 & 0 & 0 \\ q_{13} & q_{23} & q_{33} & 0 & 0 & 0 \\ q_{14} & 0 & 0 & q_{44} & q_{45} & q_{46} \\ q_{15} & 0 & 0 & q_{45} & q_{55} & q_{56} \\ q_{16} & 0 & 0 & q_{46} & q_{56} & q_{66} \end{bmatrix}$, $q_{11} = 0.95246$, $q_{12} = 3.0295 \times 10^{-5}$, $q_{13} = 1.0726 \times 10^{-3}$, $q_{14} = 8.5967 \times 10^{-2}$, $q_{15} = 2.371 \times 10^{-4}$, $q_{16} = 0.53001$, $q_{23} = -3.9912 \times 10^{-7}$, $q_{22} = 2.4374 \times 10^{-4}$, $q_{33} = 0.99998$, $q_{44} = 1.0$, $q_{45} = -1.664 \times 10^{-10}$, $q_{46} = 6.6313 \times 10^{-6}$, $q_{56} = -1.2440 \times 10^{-7}$, $q_{55} = 4.9671 \times 10^{-5}$, $q_{66} = 0.99997$; consequently, the other values are $\overline{d}^2 = 0.1^2 = 0.01$, $\alpha = \frac{1}{\theta} E^T P E \overline{d}^2 = \frac{0.16809\,(0.01)}{1 \times 10^{-2}} = 0.16809$, $\gamma = \frac{\lambda_{\max}(P)}{\lambda_{\min}(P)} = \frac{20.780}{2.4388 \times 10^{-2}} = 852.06$, $\beta = \lambda_{\min}(QP^{-1}) = 0.03158$and $\xi = \frac{\alpha}{\beta \lambda_{\min}(P)} = \frac{0.16809}{0.03158\,(2.4388 \times 10^{-2})} = 218.25$. The condition of Theorem 5.3 is satisfied with the eigenvalues of A_0 which are -0.48926, -1.8136×10^{-6}, -29.844, $-28.518 - 24.125i$, $-28.518 + 24.125i$, -1.0208×10^{-2}, where all the real parts are negative; therefore, the observer errors will converge to $\lim_{t\to\infty} |\widetilde{\chi}_1| \le 8.2016 \times 10^{-31}$, $\lim_{t\to\infty} |\widetilde{\chi}_2| \le 3.3333$, $\lim_{t\to\infty} |\widetilde{\chi}_3| \le 6.1696 \times 10^{-29}$, $\lim_{t\to\infty} |\widetilde{\chi}_4| \le 1.3666 \times 10^{-31}$, $\lim_{t\to\infty} |\widetilde{\chi}_5| \le 1.4774 \times 10^{-29}$, and $\lim_{t\to\infty} |\widetilde{\chi}_6| \le 2.4618 \times 10^{-32}$.

Figures 5.3–5.7 show the states, observer errors, input and perturbations, output, and RMSE of Observer and Observer2 applied to the wind turbine, respectively. Table 5.3 shows the RMSE of (5.17).

From Figures 5.3 and 5.4, it can be seen that observer achieves better accuracy when compared with Observer2 because the signals for the first follow better than the reference signals for the second. From Figures 5.5 and 5.6, it can be seen that

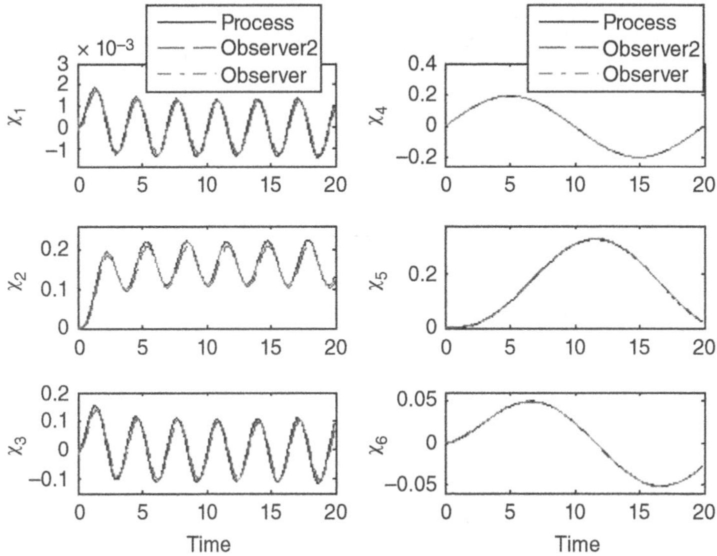

Figure 5.3 States of the wind turbine.

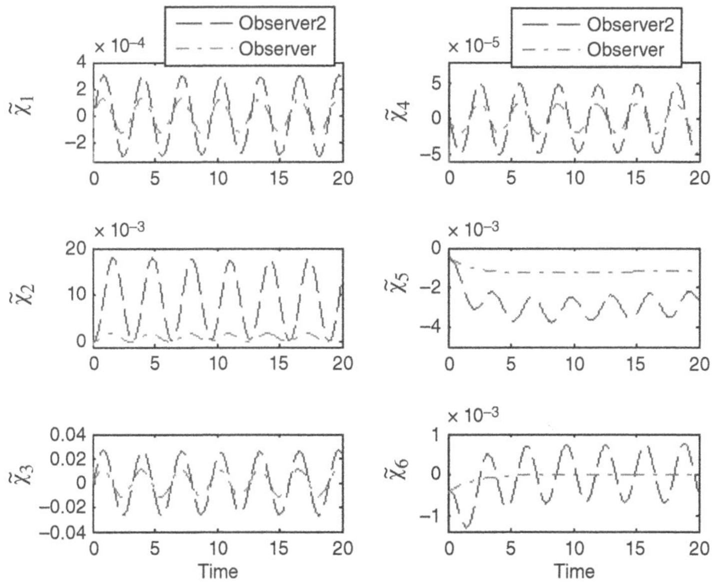

Figure 5.4 Observer errors of the wind turbine.

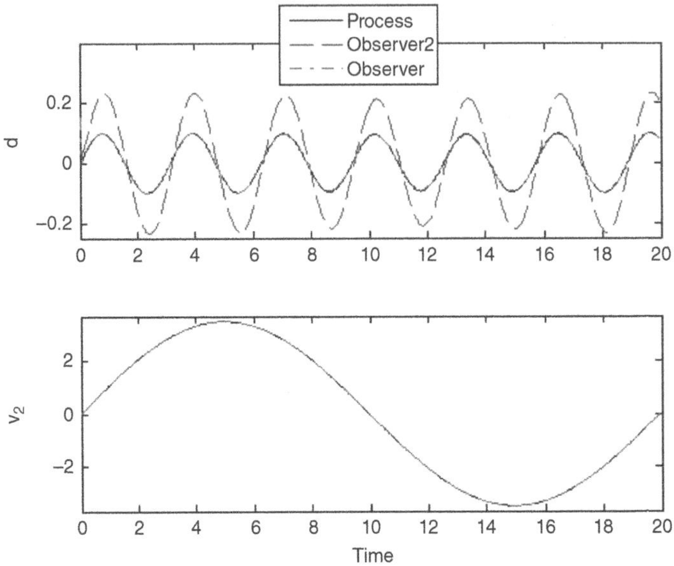

Figure 5.5 Inputs and perturbations of the wind turbine.

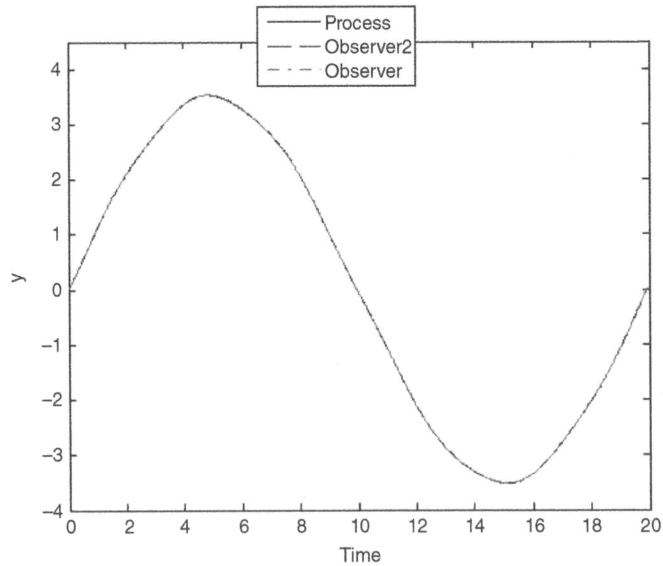

Figure 5.6 Output of the wind turbine.

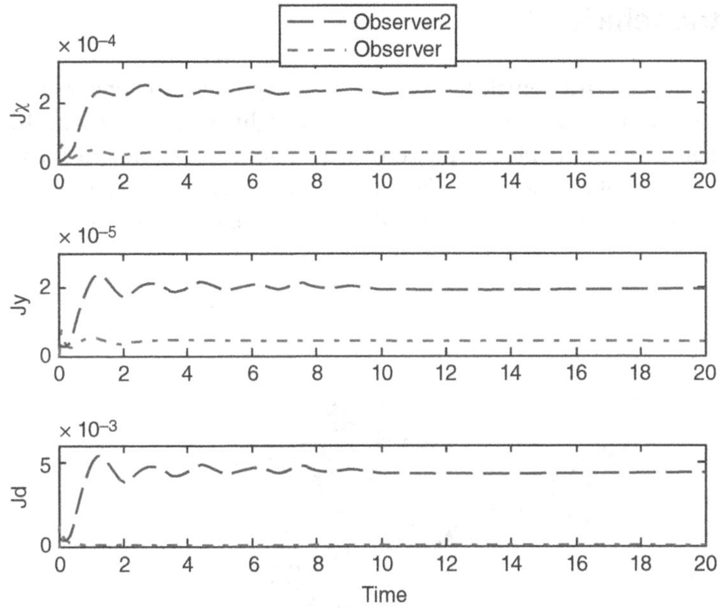

Figure 5.7 RMSE of the wind turbine.

Table 5.3 Comparisons for the wind turbine.

Strategies	Jx	Jy	Jd
Observer2	2.3581×10^{-4}	1.9804×10^{-5}	0.0042
Observer	3.3311×10^{-5}	3.6477×10^{-6}	4.7752×10^{-6}

Observer is better than Observer2 because the first follows better than the outputs of the second. From Table 5.3 and Figure 5.7, it can be observed that observer achieves better accuracy when compared with observer 2 for the state and perturbation estimation because the RMSE is smaller for the first than for the second. So, the observer obtains its benefit, the state and perturbation estimation.

Remark 5.5: The wind turbine model (5.18) is a nonlinear process because it has the term $\cos(\chi_5)$, this function is considered inside the perturbations to transform the model from a nonlinear process to a disturbed process. Later, the proposed observer is designed in the disturbed process, but it is applied in the nonlinear process. The transformations have been employed in the observer and regulator designs.

5.5 Electric Vehicle

Figure 5.8 shows the electric vehicle with energy recovery. It has two long beams, two short beams, a thin plate, eight electric motors, eight shafts, eight wheels, eight electric generators, eight plate supports of generators, and two batteries. The electric vehicle has eight independent wheels, each electric generator is mechanically connected with a wheel using a drive shaft, the wheels and generators are

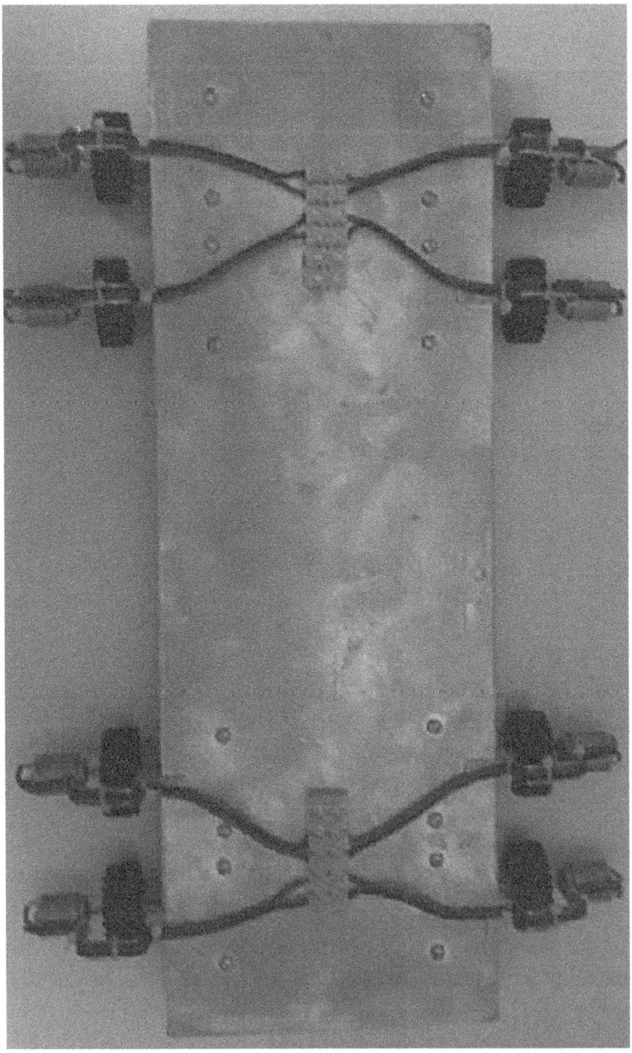

Figure 5.8 Electric vehicle.

connected with the thin plate. The wheels work at the same time and go in one direction. Batteries could be used to save the electricity taken by the electric generators or the lights could be considered to reuse the recovered energy. Some important regulators of vehicles are designed in [9, 10, 13].

The states are $\chi_1 = i_m$, $\chi_2 = v$, $\chi_3 = \omega_m$, the input is $\nu = V_{em}$, and the output is $y = V_{sg}$, the mathematical model is as follows:

$$\begin{aligned} \dot{\chi} &= A\chi + B\nu, \\ y &= C\chi + D\nu, \end{aligned} \tag{5.25}$$

where $\chi = \begin{bmatrix} \chi_1 \\ \chi_2 \\ \chi_3 \end{bmatrix}$, $A = \begin{bmatrix} -\frac{R_m}{L_m} & 0 & -\frac{K_{1m}}{L_m} \\ 0 & -\frac{8}{m}F_n\sigma_2 & \frac{8}{m}F_n\sigma_2 r \\ \frac{K_{2m}}{J_m} & 0 & -\frac{B_m}{J_m} \end{bmatrix}$, $B = \begin{bmatrix} \frac{1}{L_m} \\ 0 \\ 0 \end{bmatrix}$, $D = -C\frac{L_g}{L_m}$,

$$C = \left[c\left(\frac{L_g}{L_m}R_m - R_g\right) \;\; 0 \;\; \left(K_{1g} + K_{1m}c\frac{L_g}{L_m}\right) \right].$$

Equation (5.25) is the mathematical model of the electric vehicle with energy recovery. Table 5.4 shows the parameter descriptions for the electric vehicle. $K_{2m} = K_{aux\chi}\phi_m$. $\omega_m = \omega_g$ is considered because the electric motor and electric generator are directly connected. $i_m = \frac{1}{c}i_g$ where $c < 1$ is a parameter selected by the designer. Table 5.5 shows the parameter values for the electric vehicle with energy recovery.

The state space representation of the electric vehicle with perturbations is represented as (5.7), where d are the perturbations, ν is the regulator input, χ_1, χ_2, and χ_3 are the states, and y is the output. If the parameter values of Table 5.5 are considered, so the matrices of the process (A, B, C, D) are:

$$\begin{aligned} A &= \begin{bmatrix} -\frac{50}{0.050} & 0 & -\frac{1.537}{0.050} \\ 0 & -\frac{8(98.1)(0.01)}{0.06} & \frac{8(98.1)(0.01)(0.0225)}{0.06} \\ \frac{20}{0.04} & 0 & -\frac{0.04}{0.04} \end{bmatrix}, \\ B &= \begin{bmatrix} \frac{1}{0.050} & 0 & 0 \end{bmatrix}^T, \\ C &= \left[0.001\left(\frac{5\times 10^{-6}(50)}{0.050} - 1000\right) \;\; 0 \;\; \left(5\times 10^{-3} + \frac{1.537(0.001)(5\times 10^{-6})}{0.050}\right) \right], \\ D &= -\frac{0.001(5\times 10^{-6})}{0.050}, \end{aligned} \tag{5.26}$$

Table 5.4 Parameter descriptions for the electric vehicle.

Parameter	Description
r	Wheel radius
v, ω	Linear and angular velocity
F_n	Normal force
m	Total mass
σ_2	Viscous relative damping coefficient
i_m, i_g	Motor and generator armature currents
L_m, L_g	Motor and generator armature inductances
R_m, R_g	Motor and generator armature resistances
V_{am}	Motor armature voltage
V_{em}	Motor back emf
K_{1m}	Back emf constant
ω_m, ω_g	Motor and generator angular velocity
V_{sg}	Output generator voltage
K_{1g}	Generator torque constant
J_m	Motor inertia moment
B_m	Viscous friction coefficient
K_{2m}	Motor coefficient back emf

Table 5.5 Parameter values for the electric vehicle.

Parameter	Value	Parameter	Value
r	0.0225 m	R_m	50 Ω
σ_2	0.01 s/m	L_m	0.050 H
J_m	0.04 kg m^2	Fn	98.1 N
K_{2m}	20 kg m/A	V_{em}	24 V
K_{1m}	1.537 V/RPM	m	0.06 kg
L_g	5×10^{-6} H	R_g	1000 Ω
B_m	0.04 kg m^2/rad s	c	0.001
K_{1g}	5×10^{-3} kg m^2/rad s		

The matrix of the perturbations is:

$$E = \begin{bmatrix} \frac{1}{0.050} & 0 & 0 \end{bmatrix}^T, \tag{5.27}$$

5.5.1 Comparisons

In this section, the observer of this chapter named observer will be compared with the observer of [25] named Observer2. The benefit of the observer is that the estimated states $\hat{\chi}$ must follow the states χ and the estimated perturbations $\hat{d}$ must follow the perturbations d. The initial conditions for the plant are $\chi_0 = [1 \times 10^{-5}, 1 \times 10^{-5}, 1 \times 10^{-5}]^T$, therefore, the electric vehicle starts near the rest position. The perturbations are the undesired voltage given as $d = 0.25 \sin(20t)$V from 0 to 5 s, t is the time in s, the input voltage is $v = V_{em}$.

Observer2 of [25] is used with initial conditions $\hat{\chi}_i = [4.5 \times 10^{-4}, 4.5 \times 10^{-4}, 4.5 \times 10^{-4}]^T$ and with gains $L_i = [2 \times 10^2, 2 \times 10^1, 2 \times 10^2]^T$.

Observer of this study is employed with initial conditions $\hat{\chi}_0 = [4.5 \times 10^{-4}, 4.5 \times 10^{-4}, 4.5 \times 10^{-4}]^T$ and with gains $L = [6.6 \times 10^2, 2.8 \times 10^1, 3.8 \times 10^2]^T$. The condition of Theorem 5.2 is satisfied with $P = \begin{bmatrix} 4.7855 \times 10^{-3} & -5.3120 \times 10^{-5} & 6.2001 \times 10^{-2} \\ -5.3120 \times 10^{-5} & 3.8373 \times 10^{-3} & 1.0309 \times 10^{-4} \\ 6.2001 \times 10^{-2} & 1.0309 \times 10^{-4} & 0.85782 \end{bmatrix}$ and $Q = \begin{bmatrix} 0.99986 & -5.76 \times 10^{-7} & -1.8121 \times 10^{-4} \\ -5.76 \times 10^{-7} & 1.0 & 2.8611 \times 10^{-7} \\ -1.8121 \times 10^{-4} & 2.8611 \times 10^{-7} & 0.99998 \end{bmatrix}$; consequently, the other values are $\bar{d}^2 = 0.25^2 = 0.0625$, $\alpha = \frac{1}{\theta} E^T P E \bar{d}^2 = \frac{1.9142\,(0.0625)}{1 \times 10^{-0}} = 0.11964$, $\gamma = \frac{\lambda_{max}(P)}{\lambda_{min}(P)} = \frac{0.8623}{3.0161 \times 10^{-4}} = 2859.0$, $\beta = \lambda_{min}(QP^{-1}) = 1.1603$, and $\xi = \frac{\alpha}{\beta \lambda_{min}(P)} = \frac{0.11964}{1.1603(3.0161 \times 10^{-4})} = 341.87$. The condition of Theorem 5.3 is satisfied with the eigenvalues of A_0 which are $-1.0848, -130.8, -1658.0$, where all the real parts are negative; therefore, the observer errors will converge to $\lim_{t\to\infty} |\tilde{\chi}_1| \leq 2.5020 \times 10^{-3}$, $\lim_{t\to\infty} |\tilde{\chi}_2| \leq 8.398 \times 10^{-3}$, and $\lim_{t\to\infty} |\tilde{\chi}_3| \leq 0.33358$.

Figures 5.9–5.13 show the states, observer errors, input and perturbations, output, and RMSE of the observer and observer2 applied to the electric vehicle, respectively. Table 5.6 shows the RMSE of (5.17).

From Figures 5.9 and 5.10, it can be seen that observer achieves better accuracy when compared with Observer2 because the signals for the first follow better the reference signals than for the second. From Figures 5.11 and 5.12, it can be seen that Observer is better than Observer2 because the first follows better outputs than the second. From Table 5.3 and Figure 5.13, it can be observed that observer achieves better accuracy when compared with Observer2 for the state and perturbation estimation because the RMSE is smaller for the first than for the second. So, the observer obtains its benefit, the state and perturbation estimation.

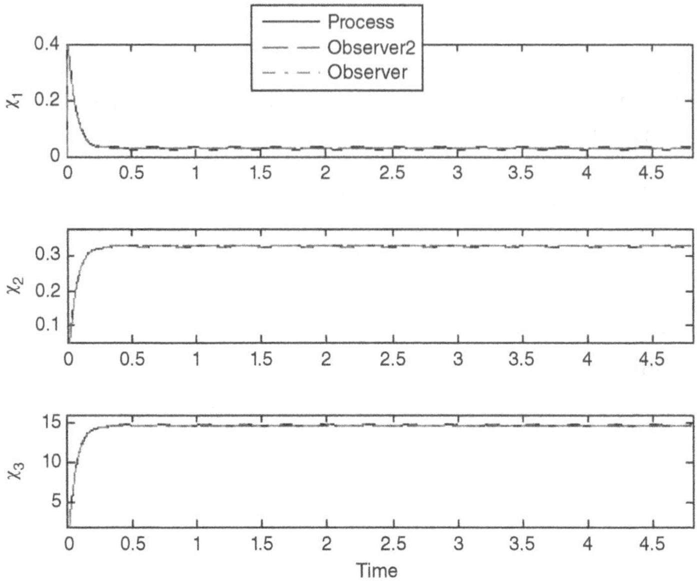

Figure 5.9 States of the electric vehicle.

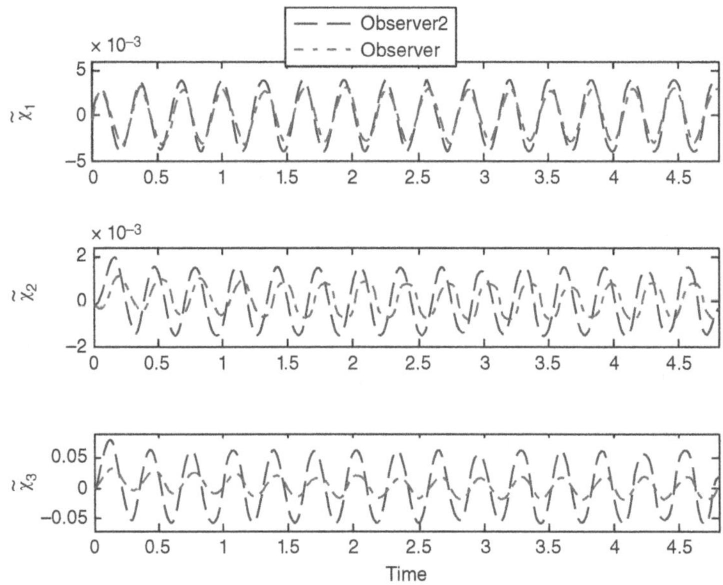

Figure 5.10 Observer errors of the electric vehicle.

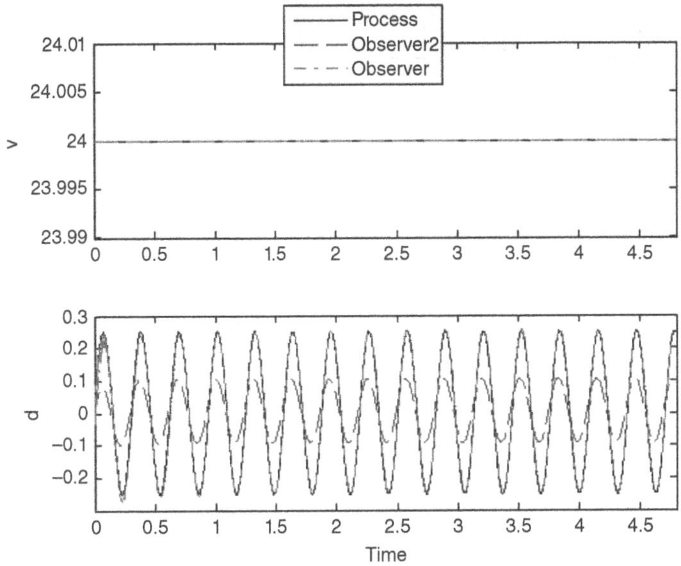

Figure 5.11 Inputs and perturbations of the electric vehicle.

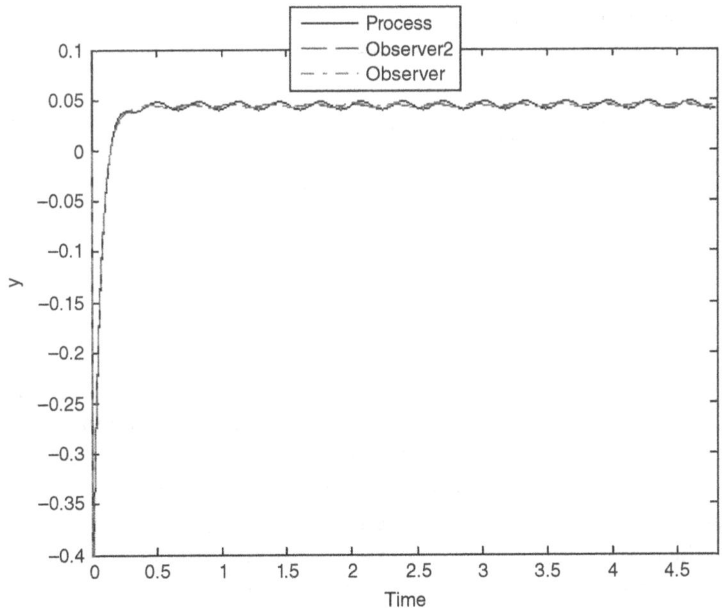

Figure 5.12 Output of the electric vehicle.

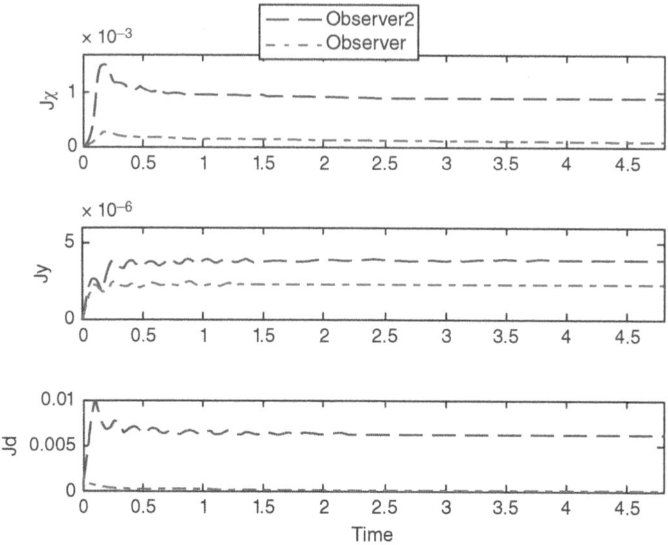

Figure 5.13 RMSE of the electric vehicle.

Table 5.6 Comparisons for the electric vehicle.

Strategies	Jx	Jy	Jd
Observer2	8.9266×10^{-4}	3.9052×10^{-6}	0.0062
Observer	9.8746×10^{-5}	2.2933×10^{-6}	4.9362×10^{-5}

5.6 Concluding Remarks

In this chapter, an observer for the state and perturbation estimation in disturbed processes was recommended. The stability of the observer was warranted by the Lyapunov method. In addition, the convergence of the observer error was warranted. The proposed observer was compared with a previous method in a wind turbine and an electric vehicle, showing that the first achieved better accuracy than the second for the state and perturbation estimation. The recommended strategy can be applied to many processes as are the electric, thermal, hydraulic, or mechanic. In the forthcoming work, other kinds of observers will be designed for the state and perturbation estimation, or the trajectory tracking problem will be considered.

References

1 Cortes-Romero J., Ramos G. A., Coral-Enriquez H. (2014), Generalized proportional integral control for periodic signals under active disturbance rejection approach, ISA Transactions, 53, 1901–1909.

2 Florin Caruntu C., Lazar C. (2014), Network delay predictive compensation based on time-delay modelling as disturbance, International Journal of Control, 87 (10), 2012–2026.

3 Nam P. T., Pathirana P. N., Trinh H. (2014), ε-bounded state estimation for time-delay systems with bounded disturbances, International Journal of Control, 87 (9), 1747–756.

4 Gao Z. (2014), On the centrality of disturbance rejection in automatic control, ISA Transactions, 53, 850–857.

5 Guo L., Cao S. (2014), Anti-disturbance control theory for systems with multiple disturbances: a survey, ISA Transactions, 53, 846–849.

6 Huang Y., Xue W. (2014), Active disturbance rejection control: methodology and theoretical analysis, ISA Transactions, 53, 963–976.

7 Laakkonen P., Pohjolainen S. (2011), Directed structure at infinity for infinite-dimensional systems, International Journal of Control, 84 (4), 702–715.

8 Liu Y., Lunze J. (2014), Leader-follower synchronisation of autonomous agents with external disturbances, International Journal of Control, 87 (9), 1914–1925.

9 Liu Y., Gao H., Xu B., Liu G., Cheng H. (2014), Autonomous coordinated control of a platoon of vehicles with multiple disturbances, IET Control Theory and Applications, 8 (18), 2325–2335.

10 Sun H., Li S., Yang J., Guo L. (2014), Non-linear disturbance observer-based back-stepping control for airbreathing hypersonic vehicles with mismatched disturbances, IET Control Theory and Applications, 8 (17), 1852–1865.

11 Wonham W. M., (1985), Linear Multivariable Control: A Geometric Approach, Springer-Verlag.

12 Wonham W. M. (1979), Geometric state-space theory in linear multivariable control: a status report, Automatica, 15 (1), 5–13.

13 Xie H., Song K., He Y. (2014), A hybrid disturbance rejection control solution for variable valve timing system of gasoline engines, ISA Transactions, 53, 889–898.

14 Yang X., Jim Zhu J., Scottedward Hodel A. (2015), Singular perturbation margin and generalised gain margin for linear time-invariant systems, International Journal of Control, 88 (1), 11–29.

15 Yao X., Zhu L., Guo L. (2014), Disturbance-observer-based control & H∞ control for non-linear Markovian jump singular systems with multiple disturbances, IET Control Theory and Applications, 8 (16), 1689–1697.

16 Yao X., Guo L., (2014), Composite disturbance-observer-based output feedback control and passive control for Markovian jump systems with multiple disturbances, IET Control Theory and Applications, 8 (10), 873–881.
17 Zhao S., Gao Z. (2014), Modified active disturbance rejection control for time-delay systems, ISA Transactions, 53, 882–888.
18 Zheng Q., Gao Z. (2014), Predictive active disturbance rejection control for processes with time delay, ISA Transactions, 53, 873–881.
19 Zhou L., She J., Wu M., He Y., Zhou S. (2014), Estimation and rejection of aperiodic disturbance in a modified repetitive-control system, IET Control Theory and Applications, 8 (10), 882–889.
20 Rubio J. J., Ochoa G., Balcazar R., Pacheco J. (2015), Uniform stable observer for the disturbance estimation in two renewable energy systems, ISA Transactions, 58, 155–164.
21 Ginter V. J., Pieper J. K. (2011), Robust gain scheduled control of a hydrokinetic turbine, IEEE Transactions on Control Systems Technology, 19 (4), 805–817.
22 Jafarnejadsani H., Pieper J., Ehlers J. (2013), Adaptive control of a variable-speed variable-pitch wind turbine using radial-basis function neural network, IEEE Transactions on Control Systems Technology, 21 (6), 2268–2272.
23 Jafarnejadsani H., Pieper J. (2015), Adaptive control of a variable-speed variable-pitch wind turbine using radial-basis function neural network, IEEE Transactions on Control Systcms Technology, 23 (1), 372–379.
24 Rubio J. J., Soriano L. A., Yu W. (2014), Dynamic model of a wind turbine for the electric energy generation, Mathematical Problems in Engineering, 2014, 1–8.
25 Rubio J. J., Melendez F., Figueroa M. (2014), An observer with controller to detect and reject disturbances, International Journal of Control, 87 (3), 524–536.

6

The Perturbation Estimation in Two Gas Processes

6.1 Introduction

The perturbations are the unwanted and unknown inlets in nonlinear processes, which can affect the outlets. This issue has occurred in many nonlinear processes. Since perturbations can affect the sensors, actuators, or processes yielding additional costs, and since most nonlinear process regulators require the knowledge of perturbations, an approach for the perturbation estimation is highly desirable.

There are some studies about regulators for perturbed processes. In [1, 2], the active strategy for the perturbation attenuation is mentioned. In [3, 4], the singular perturbations approach for the perturbation attenuation is considered. The variable and perturbation estimation in processes is focused in [5–8]. In [9–13], the robust analysis for the perturbed processes stabilization is focused. The perturbation estimation with fuzzy regulators is mentioned in [14, 15]. In [16–18], the authors use the neuro-fuzzy approximations for the perturbed processes regulation. The adaptive laws for the perturbed processes regulation are focused in [19, 20]. In [21, 22], the structure theory for the perturbation attenuation is mentioned. From the above studies, in [1, 2, 5–13, 17, 18], the authors use approaches for the variable or perturbation estimation in nonlinear processes; so, it would be welcome to be focused on this issue. The benefit of this chapter is that each nonlinear process has a different structure; so, a special estimator with the structure of the gas processes must be discussed.

There are various estimators who use the process outlets for the variable estimation [1, 2, 5, 6], but there are not many estimators who use the process outlets for the perturbation estimation. In [7, 8, 11, 13, 17, 18], previous studies of estimators are focused on the perturbation estimation, but with the two below differences: in the previous studies, estimators with adaptive or feedback outlets are used, while

Perturbations in Mechatronic, Robotic, and Energy Systems: Estimation and Attenuation, First Edition. Jose de Jesus Rubio.

in this chapter an estimator with sliding mode outlets is used; in the processes of previous studies the noise is not used, while in the processes of this chapter the noise is used.

In this chapter, the sliding mode approach is utilized in our estimator for the variable and perturbation estimation in nonlinear processes. Since the sliding mode approach uses the signum function, it can yield the unwanted chattering [3, 12, 14, 15].

The first benefit of this chapter is that an estimator is designed. It is described by the following characteristics: (1) the saturation function is used in our estimator instead of the signum function to decrease the chattering; and later, we warrant the estimator convergence by the Lyapunov strategy; (2) find the conditions required by our estimator gains that allow it to reach the variable error convergence; it yields an acceptable performance in the variable and perturbation estimation.

The second benefit of this chapter is that an algorithm is recommended to choose the gains for achieving a satisfactory performance in our estimator. It is described as follows: (1) we choose a value for gain 1; (2) we obtain the estimator matrix; (3) we obtain the eigenvalues of the estimator matrix; if the real parts of all the eigenvalues of the estimator matrix are negative, then gain 1 is correctly chosen, and we can go to step 4; otherwise, we must return to step 1; (4) we choose matrix 1 of the Lyapunov equation; (5) we substitute matrix 1 and estimator matrix into the Lyapunov equation and find matrix 2 of the Lyapunov equation; if matrices 1 and 2 of the Lyapunov equation are positive definite, then matrix 1 is correctly chosen, and we can go to step 6; otherwise, we must return to step 1; (6) we choose gains 2–4 to solve the estimator; if the estimator reaches an acceptable exactness for the variable and perturbation estimation in the nonlinear process, then the algorithm finishes; otherwise, we must return to step 6.

Our estimator is applied for the variable and perturbation estimation in the gas turbine and gasification processes. The gas turbine process is used for the electrical energy generation from the gas [23], while the gasification process is used for the gas generation from biomass [24].

The rest of the chapter is described as follows: Section 6.2 presents the estimator design containing the variable error convergence, and the perturbation estimation in nonlinear processes; later, an algorithm is recommended to choose the gains for achieving a satisfactory performance in our estimator. In Sections 6.3 and 6.4, estimators are recommended for the variable and perturbation estimation in the gas turbine and gasification processes. Section 6.5 expresses the conclusion and future work.

6.2 The Estimator for the Variable and Perturbation Estimation

In this section, a variable estimator and a perturbations estimator, which are termed as estimators, will be recommended for the variable and perturbation estimation in nonlinear processes.

In this chapter, a special nonlinear process will be used, in which the outlets have a linear combination with the variables, the variables have a nonlinear combination with the variables, and the perturbations are entered additively. The nonlinear process is [23, 24]:

$$\begin{aligned} \dot{\chi} &= A\chi + f(\chi, \nu) + Bu, \\ y &= C\chi, \end{aligned} \tag{6.1}$$

$\chi \in \Re^n$ as the process variables, $\nu \in \Re^m$ as the process inlets, $y \in \Re$ as the process outlets, $u \in \Re$ as the perturbations, $f(\chi, \nu) \in \Re^n$ as a nonlinear function, and $A \in \Re^{n\times n}, B \in \Re^{n\times 1}, C \in \Re^{p\times n}$ as matrices.

6.2.1 The Variable Estimator

The goal of the variable estimator is that, using the inlets and outlets, the variables of the variable estimator should estimate the nonlinear process variables.

The estimator error $\tilde{y}$ is:

$$\tilde{y} = y - \hat{y} = C\tilde{\chi}, \tag{6.2}$$

$\hat{y}$ as the variable estimator outlet, $\tilde{\chi} = \chi - \hat{\chi}$ as the variable error, and $\hat{\chi}$ as the estimator variables. The variable estimator is:

$$\begin{aligned} \dot{\hat{\chi}} &= A\hat{\chi} + f(\hat{\chi}, \nu) + K\tilde{y} + E\mathrm{sat}(M\tilde{y}), \\ \hat{y} &= C\hat{\chi}, \\ \mathrm{sat}(M\tilde{y}) &= \begin{cases} 1 & M\tilde{y} > 1 \\ M\tilde{y} & |M\tilde{y}| \le 1, \\ -1 & M\tilde{y} < -1 \end{cases} \end{aligned} \tag{6.3}$$

$\hat{\chi}$ as the estimator variables, $\hat{y}$ as the estimator outlets, sat$(\cdot)$ as the saturation function, and M as a matrix where $MC \in \Re^{n\times n}$ is a positive semi-definite constant, $K \in \Re^{n\times 1}$, and $E \in \Re^{n\times 1}$.

6.2.2 The Convergence Analysis of the Variable Estimator

In this section, the convergence of the variable estimator applied to nonlinear processes is analyzed based on the Lyapunov approach [3, 4].

The closed-loop model of the variable estimator is the subtraction of (6.3) to (6.1) and using the estimator error (6.2) as:

$$\begin{aligned}\dot{\tilde{\chi}} &= A\tilde{\chi} + f(\chi, \nu) + Bu - f(\hat{\chi}, \nu) \\ &-K\tilde{y} - E\text{sat}(M\tilde{y}) \\ &= A\tilde{\chi} + \tilde{f} + Bu - KC\tilde{\chi} - E\text{sat}(MC\tilde{\chi}) \\ &= A_s\tilde{\chi} + \tilde{f} + Bu - E\text{sat}(MC\tilde{\chi}),\end{aligned} \tag{6.4}$$

$\tilde{f} = f(\chi, \nu) - f(\hat{\chi}, \nu)$, $A_s = A - KC$. The nonlinear function $\tilde{f}$ is bounded as $|\tilde{f}| \leq \overline{f}$ and $|\cdot|$ as the absolute value.

Theorem 6.1 analyzes the variable estimator convergence.

Theorem 6.1: *The variable error of the variable estimator (6.2) and (6.3) applied to estimate the nonlinear process variables χ (6.1) is convergent, $\gamma = \lambda_{\min}(Q_s P_s^{-1})$, $|\tilde{f} + Bu| \leq \overline{u}$, $\overline{u} \leq E$, $\|\cdot\|$ as the Euclidean norm in $\Re^n$, $|\cdot|$ as the absolute value, and $P_s \in \Re^{n\times n}$ and $Q_s \in \Re^{n\times n}$ are positive definite matrices, which meet:*

$$A_s^T P_s + P_s A_s = -Q_s, \tag{6.5}$$

A_s as is expressed in (6.4).

Proof. See [25] for the proof. ■

6.2.3 The Perturbations Estimator

The goal of the perturbations estimator is that, using the outlets, the perturbations of the perturbations estimator should estimate the nonlinear process perturbations.

From Theorem 6.1, it is:

$$\lim_{T\to\infty} |\tilde{\chi}| \cong 0. \tag{6.6}$$

Using, $\lim_{T\to\infty} \dot{\tilde{\chi}} \leq \lim_{T\to\infty} \overline{\chi}$, $\lim_{T\to\infty} \dot{\tilde{f}} \cong 0$, $\overline{\chi}$ as the upper bound of $\tilde{\chi}$, in the first equality of (6.4) is:

$$\begin{aligned}&\lim_{T\to\infty} A\tilde{\chi} + \lim_{T\to\infty} \tilde{f} + \lim_{T\to\infty} Bu - \lim_{T\to\infty} K\tilde{y} \\ &- \lim_{T\to\infty} E\text{sat}(M\tilde{y}) = \lim_{T\to\infty} \dot{\tilde{\chi}} \leq \lim_{T\to\infty} \overline{\chi} \\ &\Rightarrow \lim_{T\to\infty} Bu - \lim_{T\to\infty} K\tilde{y} - \lim_{T\to\infty} E\text{sat}(M\tilde{y}) \leq \lim_{T\to\infty} \overline{\chi} \\ &\Rightarrow \lim_{T\to\infty} u \leq \lim_{T\to\infty} B^{-1}[K\tilde{y} + E\text{sat}(M\tilde{y}) + \overline{\chi}],\end{aligned} \tag{6.7}$$

B^{-1} as the pseudo-inverse of B. Since all the terms of (6.7) are bounded independently of T, it becomes:

$$\hat{u} \cong B^{-1}\left[K\tilde{y} + E\text{sat}(M\tilde{y}) + \overline{\chi}\right], \tag{6.8}$$

$\hat{u}$ as the perturbations of the perturbations estimator, u as the nonlinear process perturbations, and $\overline{\chi}$ as the upper bound of $\dot{\tilde{\chi}}$.

Remark 6.1: Note that the usage of the function sat(·) yields that the derivative $\dot{\tilde{\chi}}$ does not tend to zero even if the function $\tilde{\chi}$ tends to zero, but it could be said that $\dot{\tilde{\chi}}$ is bounded. This fact is used to express $\hat{u}$ as the estimated perturbations of u.

Remark 6.2: Note that in (6.2) and (6.8), $\hat{u}$ only allows to estimate perturbations u, and since in this chapter the function sat(·) is used, it can reduce the unwanted chattering. The possibility of changing the noncontinuous function sat(·) for a softer one also could reduce the chattering.

Figure 6.1 shows the variable estimator and perturbations estimator for the variable and perturbation estimation. The variable estimator of (6.2) and (6.3) and perturbations estimator of (6.2) and (6.8) are termed as the estimator of (6.2), (6.3), and (6.8).

Remark 6.3: Note that it is not required that the nonlinear process (6.1) must be convergent to achieve a satisfactory performance in our estimator.

Remark 6.4: For the satisfactory operation of the estimator (6.2), (6.3), and (6.8), the theory and application conditions mentioned must be met: (a) propose a gain K that meets the theory condition (6.5) such that the variables $\hat{\chi}$ of the variable

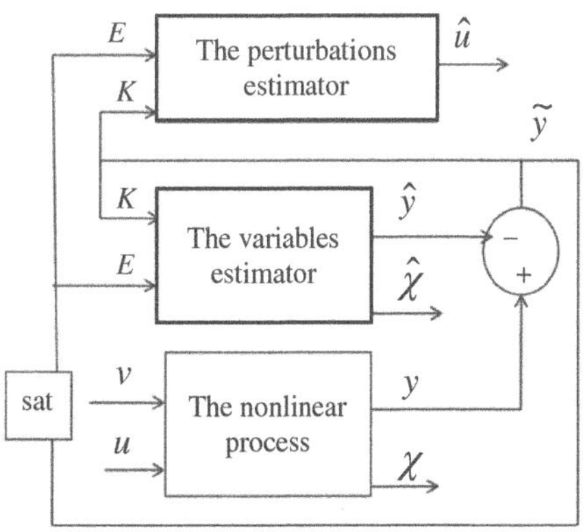

Figure 6.1 The estimators for the perturbations estimation.

estimator (6.2) and (6.3) must reach the nonlinear process variables χ of (6.1) as soon as possible; (b) propose the gain E of the perturbations estimator (6.2) and (6.8) that meets the application condition such that the estimated perturbations $\hat{u}$ of (6.2) and (6.8) should reach the nonlinear process perturbations u of (6.1) as soon as possible. In case the recommended gains K, E do not work, you have to start over.

Figure 6.2 shows the recommended algorithm to choose the gains K, E, M, $\overline{\chi}$ for achieving a satisfactory performance in our estimator. The request $\text{Re}\,(\text{eigen}(A_{ss})) < 0$ represents if the real part in the eigenvalues of A_{ss} are negative, the request $Q_s > 0$ represents if Q_s is positive definite, and the request $\hat{\chi} \cong \chi, \hat{u} \cong u$ represents if $\hat{\chi}$, $\hat{u}$ achieve a satisfactory performance in the estimation of χ, u. The recommended algorithm of Figure 6.2 is detailed as follows: (1) we choose a value for K; (2) we obtain $A_s = A - KC$; (3) we obtain the eigenvalues of $A_{ss} = sI - A_s$; if the real parts of all the eigenvalues of A_{ss} are negative, then the gain K is correctly chosen, and we can go to step 4; otherwise, we must return to step 1; (4) we choose P_s; (5) we substitute P_s and A_s into the Lyapunov equation $A_s^T P_s + P_s A_s = -Q_s$, and we find Q_s; if P_s and Q_s are positive definite, then P_s is correctly chosen; Theorem 6.1 is met, and we can go to step 6; otherwise, we must return to step 1; (6) we choose the gains E, M, and $\overline{\chi}$ to solve the recommended estimator as $\dot{\hat{\chi}} = A\hat{\chi} + f(\hat{\chi}, v) + K\tilde{y} + E\text{sat}(M\tilde{y}), \hat{y} = C\hat{\chi}$, $\hat{u} \cong B^{-1}\left[K\tilde{y} + E\text{sat}(M\tilde{y}) + \overline{\chi}\right]$, if the recommended estimator reaches an acceptable exactness for the variable and perturbation estimation in the nonlinear process $\dot{\chi} = A\chi + f(\chi, v) + Bu$, $y = C\chi$; then the algorithm finishes; otherwise, we must return to step 6.

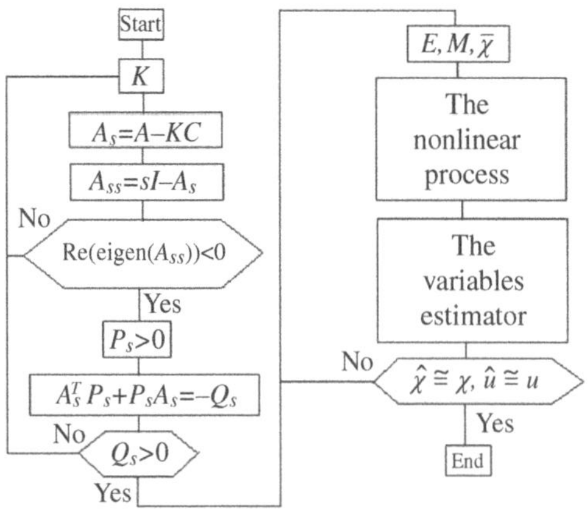

Figure 6.2 The recommended algorithm of our estimator.

In the following sections, the root mean squared error (RMSE) is used for comparisons; it is:

$$J = \left(\frac{1}{T} \int_0^T \chi^2 u\tau \right)^{\frac{1}{2}}, \tag{6.9}$$

$J = J_\chi$, $\chi^2 = \widetilde{\chi}^2$ for variables, $J = J_y$, $\chi^2 = \widetilde{y}^2$ for outlets, and $J = J_u$, $\chi^2 = \widetilde{u}^2 = (u - \widehat{u})^2$ for perturbations.

6.3 The Gas Turbine Process

Figure 6.3 shows the gas turbine process.

The gas turbine process consists of a compressor, a combustion chamber, a turbine, and a power turbine. T_3, T_4, T_5, T_6 as the temperatures in the different stages, M_3, M_4, M_5, M_6 as the amounts of mass, and m_f as the flow rate of the fuel. The inlets are $\nu_1 = m_f$, $\nu_2 = T_2$, $\nu_3 = P_2$, the variables are $\chi_1 = m_3$, $\chi_2 = m_4$, $\chi_3 = m_5$,

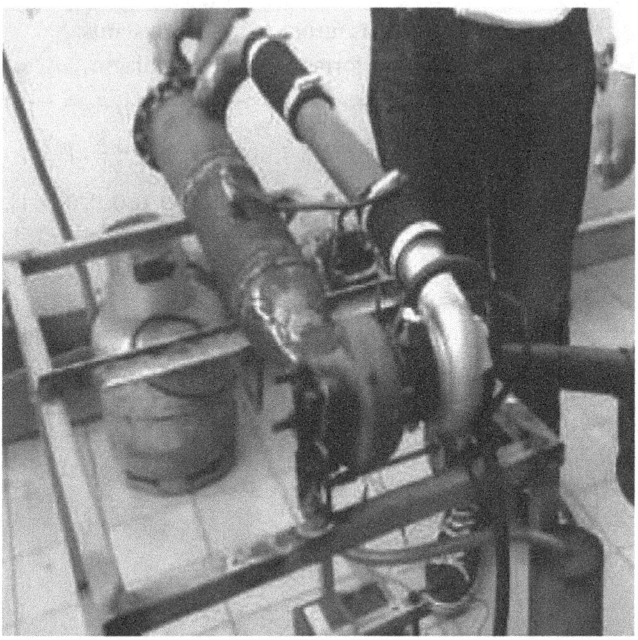

Figure 6.3 The gas turbine process.

Table 6.1 The gas turbine process constants.

Constant	Constant
$\text{MFDP} = 1 \times 10^{-2}$ kg/s	$m_a = 76$ kg/s
$m_{3\text{cor}} = 7.731$ kg/s	$T_{2\text{ref}} = 288.15$ K
$C_p = 1005$ (J/kg)K	$\eta_c = 1.46966$
$C_v = 718$ (J/kg)K	$V = 0.8$ m^3
$R = 287$ (J/kg)K	$\eta_b = 0.99$
$\text{LHV} = 44124$ (J/kg)K	$\eta_t = 1 \times 10^{-4}$
$P_{2\text{ref}} = 103.125$ Pa	$\text{DP} = 0.99$ Pa
$m_g = 32.94229$ kg/s	$\eta_{pt} = 1 \times 10^{-4}$
$P_{4\text{ref}} = 680.125$ Pa	$T_{4\text{ref}} = 1400$ K
$C_{vg} = 863$ (J/kg)K	$\gamma = 1.4$
$P_{5\text{ref}} = 416.0$ Pa	$T_{5\text{ref}} = 1000$ K
$m_{5\text{cor}} = 2.7184$ kg/s	$\sigma = 1 \times 10^{-6}$
$m_{6\text{cor}} = 0.3912$ kg/s	$\gamma_g = 1.33$

$\chi_4 = m_6$, $\chi_5 = T_3$, $\chi_6 = T_4$, $\chi_7 = T_5$, $\chi_8 = T_6$, $\chi_9 = P_3$, $\chi_{10} = P_4$, $\chi_{11} = P_5$, $\chi_{12} = P_6$, and the outlet is $y = P_6$. Table 6.1 shows the gas turbine process constants.

The gas turbine process is represented in the form (6.1) with $A = \text{diag}[a_1, a_2, a_3, a_4, a_5, a_6, a_7, a_8, a_9, a_{10}, a_{11}, a_{12}]$, $a_1 = -1$, $a_2 = -1$, $a_3 = -1$, $a_4 = -1$, $a_5 = -\frac{R}{C_v}$, $a_6 = -\frac{R}{C_{vg}}$, $a_7 = -\frac{R}{C_v}$, $a_8 = -\frac{R}{C_v}$, $a_9 = -\frac{R}{V}$, $a_{10} = -DP$, $a_{11} = -\frac{R}{V}$, $a_{12} = -\frac{R}{V}$, $f(\chi, \nu) = [f_1, f_2, f_3, f_4, f_5, f_6, f_7, f_8, f_9, f_{10}, f_{11}, f_{12}]^T$, $C = [0, 0, 0, 0, 0, 0, 0, 0, 0, 0, 0, 1]$, $\chi = [\chi_1, \chi_2, \chi_3, \chi_4, \chi_5, \chi_6, \chi_7, \chi_8, \chi_9, \chi_{10}, \chi_{11}, \chi_{12}]^T$, $\nu = [\nu_1, \nu_2, \nu_3]^T$, $B = [1, 1, 1, 1, 1, 1, 1, 1, 1, 1, 1, 1]^T$, and u as the perturbation; the terms of $f(\chi, \nu)$ are [23]:

$$f_1 = \text{MFDP} m_{3\text{cor}} \left(\frac{\nu_3 / P_{2\text{ref}}}{(\nu_2 / T_{2\text{ref}})^{0.5}} \right),$$

$$f_2 = (\chi_1 + \nu_1),$$

$$f_3 = \text{MFDP} m_{5\text{cor}} \left(\frac{\chi_{10} / P_{4\text{ref}}}{(\chi_6 / T_{4\text{ref}})^{0.5}} \right),$$

$$f_4 = \text{MFDP} m_{6\text{cor}} \left(\frac{\chi_{11} / P_{5\text{ref}}}{(\chi_7 / T_{5\text{ref}})^{0.5}} \right),$$

$$f_5 = \frac{1}{\chi_1 C_v}\left\{\text{MFDP}m_{3\text{cor}}\left(\frac{\nu_3/P_{2\text{ref}}}{(\nu_2/T_{2\text{ref}})^{0.5}}\right)\right.$$

$$\left.*\left(C_p\nu_2\left\{1+\frac{1}{\eta_c}\left[\left(\frac{\chi_9}{\nu_3}\right)^{\frac{\gamma-1}{\gamma}}-1\right]\right\}-C_v\chi_5\right)\right\},$$

$$f_6 = \frac{(C_p\chi_6 - C_{vg}\chi_6)m_a + (LHV\eta_b - C_{vg}\chi_6)\nu_1 - m_g R\chi_6 + R\chi_2\chi_6}{\chi_2 C_{vg}},$$

$$f_7 = \frac{1}{\chi_3 C_v}\left\{\text{MFDP}m_{5\text{cor}}\left(\frac{\chi_{10}/P_{4\text{ref}}}{(\chi_6/T_{4\text{ref}})^{0.5}}\right)\right.$$

$$\left.*\left(C_p\chi_6\left\{1+\eta_t\left[1-\left(\frac{\chi_{11}}{\chi_{10}}\right)^{\frac{\gamma_g-1}{\gamma_g}}\right]\right\}-C_v\chi_7\right)\right\},$$

$$f_8 = \frac{1}{\chi_4 C_v}\left\{\text{MFDP}m_{6\text{cor}}\left(\frac{\chi_{11}/P_{5\text{ref}}}{(\chi_7/T_{5\text{ref}})^{0.5}}\right)\right.$$

$$\left.*\left(C_p\chi_7\left\{1+\eta_{pt}\left[1-\left(\frac{\chi_{12}}{\chi_{11}}\right)^{\frac{\gamma_g-1}{\gamma_g}}\right]\right\}-C_v\chi_8\right)\right\},$$

$$f_9 = \frac{R}{V}\left(\chi_9 + \chi_1\dot{\chi}_5 + \chi_5\dot{\chi}_1\right),$$

$$f_{10} = DP\left(\chi_{10} + \dot{\chi}_9\right),$$

$$f_{11} = \frac{R}{V}\left(\chi_{11} + \chi_3\dot{\chi}_7 + \chi_7\dot{\chi}_3\right),$$

$$f_{12} = \frac{R}{V}\left(\chi_{12} + \chi_4\dot{\chi}_8 + \chi_8\dot{\chi}_4\right).$$

6.3.1 Comparisons

In this section, the estimator of this research named Estimator 1 is compared against the estimator of [7, 8], named Estimator 2. The initial conditions of the process are $\chi_0 = [75.9, 75.9, 75.9, 75.9, 691.7694, 1.6164\times10^3, 1.2708\times10^3, 932.4205, 2.0561\times10^6, 2.0355\times10^6, 5.7420\times10^5, 1.2679\times10^5]^T$. The goal of the estimators is that the variables $\hat{\chi}$ of the variable estimator have to reach the process variables χ and that the perturbations $\hat{u}$ of the perturbations estimator have to reach the process perturbations u. The process inlets are $\nu_1 = 1.6 + 0.1\sin(2T)$ kg/s, $\nu_2 = 288 + 1\sin(2T)$ K, and $\nu_3 = 1.013\times10^5 + 1\sin(2T)$ Pa from 0 s to 20 s. The perturbation is $u = 7.6923\times10^{-3}\sin(2T)$ from 0 to 20 s.

Estimator 2 is expressed as [7, 8] with initial conditions as $\hat{\chi}_0 = [76, 76, 76, 76, 691, 1.616\times10^3, 1.27\times10^3, 932, 2.05\times10^6, 2.03\times10^6, 5.74\times10^5, 1.24\times10^5]^T$ and gains as $L = [4\times10^{-8}, 4\times10^{-8}, 2\times10^{-8}, 2\times10^{-8}, 2\times10^{-8}, 4\times10^{-8}, 4\times10^{-8}, 4\times10^{-8}, 6\times10^{-6}, 6\times10^{-6}, 6\times10^{-6}, 6\times10^{-6}]^T$.

Estimator 1 is expressed as (6.2), (6.3), and (6.8) with $\hat{\chi} = [\hat{\chi}_1, \hat{\chi}_2, \hat{\chi}_3, \hat{\chi}_4, \hat{\chi}_5, \hat{\chi}_6, \hat{\chi}_7, \hat{\chi}_8, \hat{\chi}_9, \hat{\chi}_{10}, \hat{\chi}_{11}, \hat{\chi}_{12}]^T$, $f(\hat{\chi}, \nu) = [\hat{f}_1, \hat{f}_2, \hat{f}_3, \hat{f}_4, \hat{f}_5, \hat{f}_6, \hat{f}_7, \hat{f}_8, \hat{f}_9, \hat{f}_{10}, \hat{f}_{11}, \hat{f}_{12}]^T$; the terms of $f(\hat{\chi}, \nu)$ are:

$$\hat{f}_1 = \mathrm{MFDP}m_{3\mathrm{cor}}\left(\frac{\nu_3/P_{2\mathrm{ref}}}{(\nu_2/T_{2\mathrm{ref}})^{0.5}}\right),$$

$$\hat{f}_2 = (\hat{\chi}_1 + \nu_1),$$

$$\hat{f}_3 = \mathrm{MFDP}m_{5\mathrm{cor}}\left(\frac{\hat{\chi}_{10}/P_{4\mathrm{ref}}}{(\hat{\chi}_6/T_{4\mathrm{ref}})^{0.5}}\right),$$

$$\hat{f}_4 = \mathrm{MFDP}m_{6\mathrm{cor}}\left(\frac{\hat{\chi}_{11}/P_{5\mathrm{ref}}}{(\hat{\chi}_7/T_{5\mathrm{ref}})^{0.5}}\right),$$

$$\hat{f}_5 = \frac{1}{\hat{\chi}_1 C_v}\left\{\mathrm{MFDP}m_{3\mathrm{cor}}\left(\frac{\nu_3/P_{2\mathrm{ref}}}{(\nu_2/T_{2\mathrm{ref}})^{0.5}}\right) * \left(C_p\nu_2\left\{1+\frac{1}{\eta_c}\left[\left(\frac{\hat{\chi}_9}{\nu_3}\right)^{\frac{\gamma-1}{\gamma}} - 1\right]\right\} - C_v\hat{\chi}_5\right)\right\},$$

$$\hat{f}_6 = \frac{(C_p\hat{\chi}_6 - C_{vg}\hat{\chi}_6)m_a + (LHV\eta_b - C_{vg}\hat{\chi}_6)\nu_1 - m_g R\hat{\chi}_6 + R\hat{\chi}_2\hat{\chi}_6}{\hat{\chi}_2 C_{vg}},$$

$$\hat{f}_7 = \frac{1}{\hat{\chi}_3 C_v}\left\{\mathrm{MFDP}m_{5\mathrm{cor}}\left(\frac{\hat{\chi}_{10}/P_{4\mathrm{ref}}}{(\hat{\chi}_6/T_{4\mathrm{ref}})^{0.5}}\right) * \left(C_p\hat{\chi}_6\left\{1+\eta_t\left[1-\left(\frac{\hat{\chi}_{11}}{\hat{\chi}_{10}}\right)^{\frac{\gamma_g-1}{\gamma_g}}\right]\right\} - C_v\hat{\chi}_7\right)\right\},$$

$$\hat{f}_8 = \frac{1}{\hat{\chi}_4 C_v}\left\{\mathrm{MFDP}m_{6\mathrm{cor}}\left(\frac{\hat{\chi}_{11}/P_{5\mathrm{ref}}}{(\hat{\chi}_7/T_{5\mathrm{ref}})^{0.5}}\right) * \left(C_p\hat{\chi}_7\left\{1+\eta_{pt}\left[1-\left(\frac{\hat{\chi}_{12}}{\hat{\chi}_{11}}\right)^{\frac{\gamma_g-1}{\gamma_g}}\right]\right\} - C_v\hat{\chi}_8\right)\right\},$$

$$\hat{f}_9 = \frac{R}{V}\left(\hat{\chi}_9 + \hat{\chi}_1\dot{\hat{\chi}}_5 + \hat{\chi}_5\dot{\hat{\chi}}_1\right),$$

$$\hat{f}_{10} = DP\left(\hat{\chi}_{10} + \dot{\hat{\chi}}_9\right),$$

$$\hat{f}_{11} = \frac{R}{V}\left(\hat{\chi}_{11} + \hat{\chi}_3\dot{\hat{\chi}}_7 + \hat{\chi}_7\dot{\hat{\chi}}_3\right),$$

$$\hat{f}_{12} = \frac{R}{V}\left(\hat{\chi}_{12} + \hat{\chi}_4\dot{\hat{\chi}}_8 + \hat{\chi}_8\dot{\hat{\chi}}_4\right).$$

The initial conditions are $\hat{\chi}_0 = [76, 76, 76, 76, 691, 1.616\times 10^3, 1.27\times 10^3, 932, 2.05\times 10^6, 2.03\times 10^6, 5.74\times 10^5, 1.24\times 10^5]^T$.

The gains $K \in \Re^{12}$, $E \in \Re^{12}$, $M \in \Re^{12}$, and $\overline{\chi} \in \Re$ for the Estimator 1 are chosen using the recommended algorithm of Figure 6.2 is detailed as follows: (1) we choose a value for $K = [4\times 10^{-9}, 4\times 10^{-9}, 2\times 10^{-9}, 2\times 10^{-9}, 2\times 10^{-9}, 4\times 10^{-9}, 4\times 10^{-9}, 4\times 10^{-9}, 6\times 10^{-7}, 6\times 10^{-7}, 6\times 10^{-7}, 6\times 10^{-7}]^T$; (2) we obtain $A_s = A - KC = [a_{sij}] \in \Re^{12\times 12}$, $a_{s11} = -1$, $a_{s22} = -1$, $a_{s33} = -1$, $a_{s44} = -1$, $a_{s55} = -\frac{287}{718}$, $a_{s66} = -\frac{287}{863}$, $a_{s77} = -\frac{287}{718}$, $a_{s88} = -\frac{287}{718}$, $a_{s99} = -358.75$, $a_{s1010} = -0.99$, $a_{s1111} = -358.75$, $a_{s1212} = -358.75$, $a_{s112} = -\frac{1}{500000000}$, $a_{s212} = -\frac{1}{500000000}$, $a_{s312} = -\frac{1}{500000000}$, $a_{s412} = -\frac{1}{500000000}$, $a_{s512} = -\frac{1}{500000000}$, $a_{s612} = -\frac{1}{500000000}$, $a_{s712} = -\frac{1}{500000000}$, $a_{s812} = -\frac{1}{500000000}$, $a_{s912} = -\frac{3}{500000000}$, $a_{s1012} = -\frac{3}{500000000}$, $a_{s1112} = -\frac{3}{500000000}$; the other terms of A_s have a value of 0; (3) we obtain the eigenvalues of $A_{ss} = sI - A_s$ as $-0.42028 + 0.11615i$, $-0.42028 - 0.11615i$, -347.89, -1.3013, -18.240, $-364.18 - 9.7741i$, $-364.18 + 9.7741i$, $-0.31231 - 4.1895\times 10^{-2}i$, $-0.31231 + 4.1895\times 10^{-2}i$, $-0.75333 + 0.25701i$, $-0.75333 - 0.25701i$, -1.124; if the real parts of all the eigenvalues of A_{ss} are negative, then the gain $K \in \Re^{12}$ is correctly chosen, and we can go to step 4; otherwise, we must return to step 1, (4) we choose $P_s = [p_{ij}] \in \Re^{12\times 12}$, $p_{11} = 0.5$, $p_{22} = 0.5$, $p_{33} = 0.5$, $p_{44} = 0.5$, $p_{55} = 1.2509$, $p_{66} = 1.5035$, $p_{77} = 1.2509$, $p_{88} = 1.2509$, $p_{99} = 1.3937\times 10^{-3}$, $p_{1010} = 0.50505$, $p_{1111} = 1.3937\times 10^{-3}$, $p_{1212} = 1.3937\times 10^{-3}$; the other terms of P_s have a value of 0; (5) we substitute $P_s = [p_{ij}] \in \Re^{12\times 12}$ and $A_s = A - KC$ into the Lyapunov equation $A_s^T P_s + P_s A_s = -Q_s$, and we find $Q_s = [q_{ij}] \in \Re^{12\times 12}$, $q_{11} = 1.0$, $q_{22} = 1.0$, $q_{33} = 1.0$, $q_{44} = 1.0$, $q_{55} = 1.0$, $q_{66} = 1.0$, $q_{77} = 1.0$, $q_{88} = 1.0$, $q_{99} = 0.99998$, $q_{1010} = 1.0$, $q_{1111} = 0.99998$, $q_{1212} = 0.99998$; the other terms of Q_s have a value of 0; if $P_s = [p_{ij}] \in \Re^{12\times 12}$ and $Q_s = [q_{ij}] \in \Re^{12\times 12}$ are positive definite, then $P_s = [p_{ij}] \in \Re^{12\times 12}$ is correctly chosen, Theorem 6.1 is met, and we can go to step 6; otherwise, we must return to step 1; (6) we choose the gains $E = [1\times 10^{-10}, 1\times 10^{-10}, 1\times 10^{-10}, 1\times 10^{-10}, 1\times 10^{-10}, 1\times 10^{-10}, 1\times 10^{-10}, 1\times 10^{-10}, 1\times 10^{-10}, 1\times 10^{-10}, 1\times 10^{-10},$

$1 \times 10^{-10}]^T$, $M = [1, 1, 1, 1, 1, 1, 1, 1, 1, 1, 1, 1]^T$, and $\overline{\chi} = 2.14$ to solve the Estimator 1 as $\dot{\widehat{\chi}} = A\widehat{\chi} + f(\widehat{\chi}, \nu) + K\widetilde{y} + E\text{sat}(M\widetilde{y})$, $\widehat{y} = C\widehat{\chi}$, $\widehat{u} \cong B^{-1}[K\widetilde{y} + E\text{sat}(M\widetilde{y}) + \overline{\chi}]$, if the Estimator 1 reaches an acceptable exactness for the variable and perturbation estimation in the nonlinear process $\dot{\chi} = A\chi + f(\chi, \nu) + Bu, y = C\chi$; then the algorithm finishes; otherwise, we must return to step 6.

Figures 6.4 and 6.5 show the variable and perturbation estimation of the Estimators 1 and 2 applied to the gas turbine process. Table 6.2 shows the MSE of (6.9).

In Figures 6.4 and 6.5, since the Estimator 1 reaches better variable and perturbation estimation of the process than the Estimator 2, it is seen that the Estimator 1 reaches better performance. In addition, Figures 6.4 and 6.5 show that in the variable and perturbation estimation of Estimator 1, the unwanted chattering is not presented. In Table 6.2, since the MSE is smaller for the Estimator 1 than for the Estimator 2, it is seen that the Estimator 1 reaches better exactness for the variable and perturbation estimation.

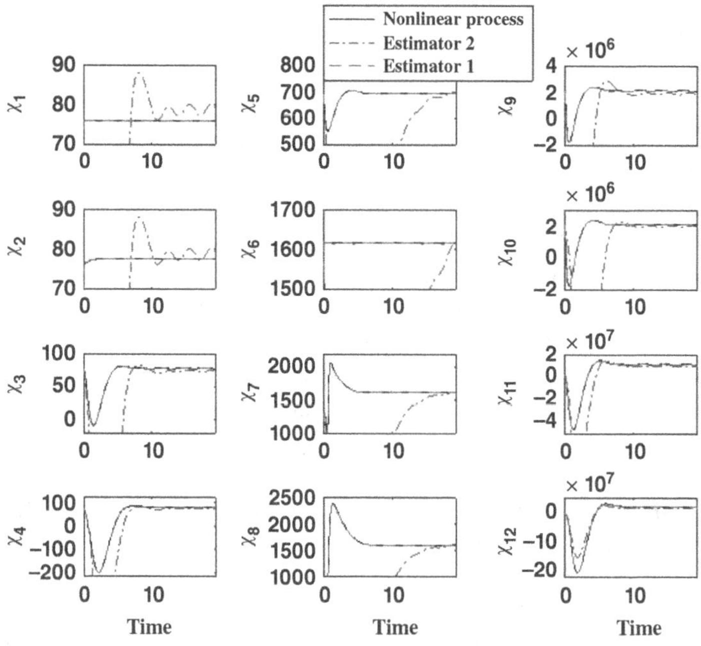

Figure 6.4 The variables estimation in the gas turbine process.

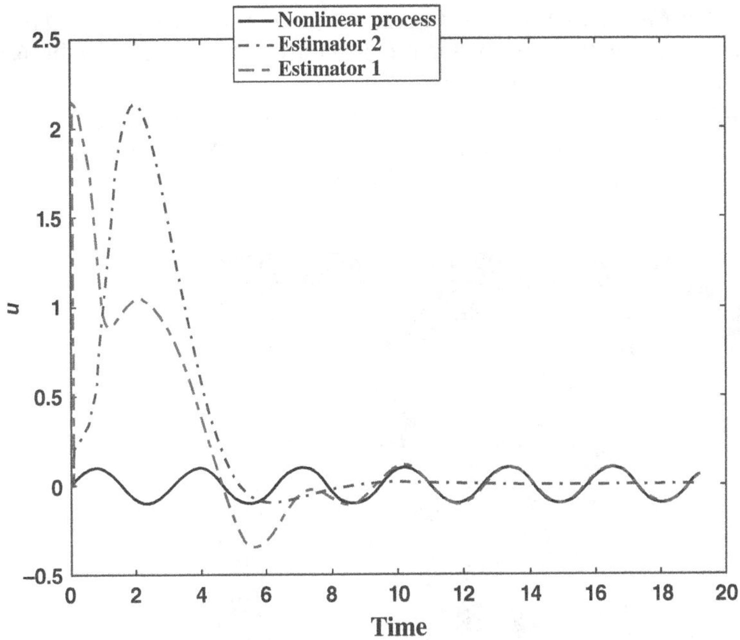

Figure 6.5 The perturbations estimation in the gas turbine process.

Table 6.2 The gas turbine process comparisons.

	Estimator 2	Estimator 1
J_χ	4.4084	0.0255
J_y	3.2915	0.0248
J_u	0.1134	0.0387

6.4 The Gasification Process

Figure 6.6 shows the gasification process.

The gasification process model consists of drying, pyrolysis, oxidation or combustion, and reduction. C_{CO}, C_{CO_2}, C_{H_2}, C_{H_4}, C_{Tar}, C_{H_2O}, C_{O_2}, C_{N_2} as concentrations of gases CO, CO_2, H_2, H_4, Tar, H_2O, O_2, N_2 in mol/cm^3, ρ_{coal}, ρ_{char} as the coal solids densities, char in g/cm^3, T as the gases temperature in K, T_s as the solids temperature in K, u as the injected gases flow in mols/cm^2/s, δ as the steam flow in mols/cm^2/s, R_1, R_2, R_3 as the chemical reactions, and m_{O_2}, m_{H_2O} as the internal molar fractions of O_2 and H_2O. The variables are $\chi_1 = \rho_{coal}$, $\chi_2 = \rho_{char}$, $\chi_3 = T_s$,

$\chi_4 = C_{CO}$, $\chi_5 = C_{CO_2}$, $\chi_6 = C_{H_2}$, $\chi_7 = C_{H_4}$, $\chi_8 = C_{Tar}$, $\chi_9 = C_{H_2O}$, $\chi_{10} = C_{O_2}$, $\chi_{11} = C_{N_2}$, the inlet is $\nu = u$, the perturbation is δ, and the outlet is $y = C_{H_2O}$. Table 6.3 shows the gasification process constants.

The gasification process has the form of (6.1) with $\chi = [\chi_1, \chi_2, \chi_3, \chi_4, \chi_5, \chi_6, \chi_7, \chi_8, \chi_9, \chi_{10}, \chi_{11}]^T$, $\nu = u$, $A = -\beta\text{diag}(1) \in \Re^{11\times 11}$, diag(1) as a diagonal matrix with a number 1 in the main diagonal, $C = [0, 0, 0, 0, 0, 0, 0, 0, 1, 0, 0]$, $B = [0, 0,$

Figure 6.6 The gasification process.

$0, 0, 0, 0, 0, 0, \frac{1}{L}, 0, 0]^T$, $f(\chi, \nu) = [f_1, f_2, f_3, f_4, f_5, f_6, f_7, f_8, f_9, f_{10}, f_{11}]^T$; the terms of $f(\chi, \nu)$ are [24]:

$$
\begin{aligned}
f_1 &= -M_1 R_1, \\
f_2 &= M_2 \left[p_{R_1 \text{char}} R_1 - R_2 - R_3 \right], \\
f_3 &= \frac{1}{C_s} [h_t (T - \chi_3) - \Delta q_2 R_2 - \Delta q_3 R_3], \\
f_4 &= p_{R_1 \text{CO}} R_1 + R_3, \\
f_5 &= p_{R_1 \text{CO}_2} R_1 + R_2, \\
f_6 &= p_{R_1 \text{H}_2} R_1 + R_3, \\
f_7 &= p_{R_1 \text{CH}_4} R_1, \\
f_8 &= p_{R_1 \text{Tar}} R_1, \\
f_9 &= p_{R_1 \text{H}_2\text{O}} R_1 + p_{R_2 \text{H}_2\text{O}} R_2 - p_{R_3 \text{H}_2\text{O}} R_3 + \frac{\Pi}{L} \nu, \\
f_{10} &= -p_{R_2 \text{O}_2} R_2 + \frac{\Xi}{L} \nu, \\
f_{11} &= \frac{\Psi}{L} \nu, \\
R_1 &= 5 \frac{\chi_1}{M_1} \exp\left(\frac{-6039}{\chi_3} \right),
\end{aligned}
$$

Table 6.3 The gasification process constants.

Constant	Constant
$p_{R_1 \text{char}} = 0.766$	$C_s = 7.3920\,\text{cal/g/K}$
$p_{R_1 \text{CO}} = 0.008$	$h_t = 0.001\,\text{cal/s/K/cm}^3$
$p_{R_1 \text{CO}_2} = 0.058$	$\beta = 7 \times 10^{-6}\,\text{mol/cm}^3\text{/s}$
$p_{R_1 \text{H}_2} = 0.083$	$\Delta q_2 = -93929\,\text{cal/mol}$
$p_{R_1 \text{CH}_4} = 0.044$	$\Delta q_3 = 31309.7\,\text{cal/mol}$
$p_{R_1 \text{Tar}} = 0.0138$	$\Pi = 0.77$
$p_{R_1 \text{H}_2\text{O}} = 0.055$	$\Xi = 0.154$
$p_{R_2 \text{H}_2\text{O}} = 0.075$	$\Psi = 7.6$
$p_{R_3 \text{H}_2\text{O}} = 0.925$	$M_1 = 54.9916\,\text{g/mol}$
$p_{R_2 \text{O}_2} = 1.02$	$M_2 = 52.9666\,\text{g/mol}$
$P = 4.83$	$T = 430\,\text{K}$
$L = 100\,\text{cm}$	

$$R_{m2} = \frac{1}{10} h_t m_{O_2},$$

$$R_{c2} = \frac{1}{M_2}\left[9.55 \times 10^8 \chi_2 m_{O_2} P \exp\left(\frac{-22142}{\chi_3}\right)\chi_3^{-0.5}\right],$$

$$R_2 = \frac{1}{\frac{1}{R_{c2}} + \frac{1}{R_{m2}}},$$

$$R_{m3} = \frac{1}{10} h_t m_{H_2O},$$

$$R_{c3} = \frac{\chi_2 m_{H_2O}^2 P^2 \exp\left(5.052 - \frac{12908}{\chi_3}\right)}{M_2\left[m_{H_2O}P + \exp\left(-22.216 + \frac{24880}{\chi_3}\right)\right]^2},$$

$$R_3 = \frac{1}{\frac{1}{R_{c3}} + \frac{1}{R_{m3}}},$$

$$m_{O_2} = \frac{\chi_5}{C_T + \chi_9},$$

$$m_{H_2O} = \frac{\chi_9}{C_T + \chi_9},$$

$$C_T = \chi_4 + \chi_5 + \chi_6 + \chi_7 + \chi_8 + \chi_{10} + \chi_{11}.$$

6.4.1 Comparisons

In this section, the estimator of this research named Estimator 1 is compared against the estimator of [7, 8], named Estimator 2. The process initial conditions are $\chi_0 = [0.48,\ 0,\ 480,\ 0,\ 0,\ 0,\ 0,\ 0,\ 0,\ 4.2 \times 10^{-4},\ 1.6 \times 10^{-6}]^T$. The goal of the estimators is that the variables $\hat{\chi}$ of the variable estimator have to reach the process variables χ and that the perturbations $\hat{u}$ of the perturbations estimator have to reach the process perturbations u. The process inlet is $\nu = 0.2 \times 10^{-3}\sin(2T)$ moles/cm^2/s from 0 s to 20 s. The perturbation is $u = 1.5 \times 10^{-5}\sin(2T)$ moles/cm^2/s from 0 s to 20 s.

Estimator 2 is expressed as [7, 8] with initial conditions as $\hat{\chi}_0 = [0.48, 6 \times 10^0, 480, 1 \times 10^{-9}, 1 \times 10^{-8}, 1 \times 10^{-8}, 1 \times 10^{-8}, 1 \times 10^{-9}, 1 \times 10^{-7}, 4.2 \times 10^4, 1.6 \times 10^{-6}]^T$ and gains as $L = [1 \times 10^{-1}, 1 \times 10^1, 1 \times 10^{-1}, 2 \times 10^{-3}, 1 \times 10^{-2}, 2 \times 10^{-2}, 1 \times 10^{-2}, 4 \times 10^{-3}, 1 \times 10^{-2}, 1 \times 10^{-3}, 1 \times 10^{-1}]^T$.

Estimator 1 is expressed as (6.2), (6.3), and (6.8) with $\hat{\chi} = [\hat{\chi}_1, \hat{\chi}_2, \hat{\chi}_3, \hat{\chi}_4, \hat{\chi}_5, \hat{\chi}_6, \hat{\chi}_7, \hat{\chi}_8, \hat{\chi}_9, \hat{\chi}_{10}, \hat{\chi}_{11}]^T$, $f(\hat{\chi}, \nu) = [\hat{f}_1, \hat{f}_2, \hat{f}_3, \hat{f}_4, \hat{f}_5, \hat{f}_6, \hat{f}_7, \hat{f}_8, \hat{f}_9, \hat{f}_{10}, \hat{f}_{11}]^T$; the terms of $f(\hat{\chi}, \nu)$ are:

$$\hat{f}_1 = -M_1\hat{R}_1,$$

$$\hat{f}_2 = M_2\left[p_{R_1\text{char}}\hat{R}_1 - \hat{R}_2 - \hat{R}_3\right],$$

$$\widehat{f}_3 = \frac{1}{C_s}\left[h_t(T - \widehat{\chi}_3) - \Delta q_2\widehat{R}_2 - \Delta q_3\widehat{R}_3\right],$$

$$\widehat{f}_4 = p_{R_1\mathrm{CO}}\widehat{R}_1 + \widehat{R}_3,$$

$$\widehat{f}_5 = p_{R_1\mathrm{CO}_2}\widehat{R}_1 + \widehat{R}_2,$$

$$\widehat{f}_6 = p_{R_1\mathrm{H}_2}\widehat{R}_1 + \widehat{R}_3,$$

$$\widehat{f}_7 = p_{R_1\mathrm{CH}_4}\widehat{R}_1,$$

$$\widehat{f}_8 = p_{R_1\mathrm{Tar}}\widehat{R}_1,$$

$$\widehat{f}_9 = p_{R_1\mathrm{H}_2\mathrm{O}}\widehat{R}_1 + p_{R_2\mathrm{H}_2\mathrm{O}}\widehat{R}_2 - p_{R_3\mathrm{H}_2\mathrm{O}}\widehat{R}_3 + \frac{\Pi}{L}\nu,$$

$$\widehat{f}_{10} = -p_{R_2\mathrm{O}_2}\widehat{R}_2 + \frac{\Xi}{L}\nu,$$

$$\widehat{f}_{11} = \frac{\Psi}{L}\nu,$$

$$\widehat{R}_1 = 5\frac{\widehat{\chi}_1}{M_1}\exp\left(\frac{-6039}{\widehat{\chi}_3}\right),$$

$$\widehat{R}_{m2} = \frac{1}{10}h_t\widehat{m}_{\mathrm{O}_2},$$

$$\widehat{R}_{c2} = \frac{1}{M_2}\left[9.55\times 10^8\widehat{\chi}_2 m_{\mathrm{O}_2}P\exp\left(\frac{-22142}{\widehat{\chi}_3}\right)\widehat{\chi}_3^{-0.5}\right],$$

$$\widehat{R}_2 = \frac{1}{\frac{1}{\widehat{R}_{c2}} + \frac{1}{\widehat{R}_{m2}}},$$

$$\widehat{R}_{m3} = \frac{1}{10}h_t\widehat{m}_{\mathrm{H}_2\mathrm{O}},$$

$$\widehat{R}_{c3} = \frac{\widehat{\chi}_2\widehat{m}^2_{\mathrm{H}_2\mathrm{O}}P^2\exp\left(5.052 - \frac{12908}{\widehat{\chi}_3}\right)}{M_2\left[\widehat{m}_{\mathrm{H}_2\mathrm{O}}P + \exp\left(-22.216 + \frac{24880}{\widehat{\chi}_3}\right)\right]^2},$$

$$\widehat{R}_3 = \frac{1}{\frac{1}{\widehat{R}_{c3}} + \frac{1}{\widehat{R}_{m3}}},$$

$$\widehat{m}_{\mathrm{O}_2} = \frac{\widehat{\chi}_5}{\widehat{C}_T + \widehat{\chi}_9},$$

$$\widehat{m}_{\mathrm{H}_2\mathrm{O}} = \frac{\widehat{\chi}_9}{\widehat{C}_T + \widehat{\chi}_9},$$

$$\widehat{C}_T = \widehat{\chi}_4 + \widehat{\chi}_5 + \widehat{\chi}_6 + \widehat{\chi}_7 + \widehat{\chi}_8 + \widehat{\chi}_{10} + \widehat{\chi}_{11}.$$

The initial conditions as $\hat{\chi}_0 = [0.48, 1 \times 10^{-5}, 480, 1 \times 10^{-9}, 1 \times 10^{-8}, 1 \times 10^{-8}, 1 \times 10^{-8}, 1 \times 10^{-9}, 1 \times 10^{-7}, 4.2 \times 10^4, 1.6 \times 10^{-6}]^T$.

The gains $K \in \Re^{11}, E \in \Re^{11}, M \in \Re^{11}$, and $\overline{\chi} \in \Re$ for the Estimator 1 are chosen using the recommended algorithm of Figure 6.2 is detailed as follows: (1) we choose a value for $K = [1 \times 10^{-1}, 1 \times 10^{0}, 1 \times 10^{-1}, 1 \times 10^{-4}, 1 \times 10^{-4}, 1 \times 10^{-4}, 1 \times 10^{-4}, 1 \times 10^{-4}, 6 \times 10^{-1}, 1 \times 10^{-2}, 1 \times 10^{-1}]^T$; (2) we obtain $A_s = A - KC = [a_{sij}] \in \Re^{11\times11}$, $a_{s11} = -\frac{7}{1000000}$, $a_{s22} = -\frac{7}{1000000}$, $a_{s33} = -\frac{7}{1000000}$, $a_{s44} = -\frac{7}{1000000}$, $a_{s55} = -\frac{7}{1000000}$, $a_{s66} = -\frac{7}{1000000}$, $a_{s77} = -\frac{7}{1000000}$, $a_{s88} = -\frac{7}{1000000}$, $a_{s99} = -\frac{600007}{1000000}$, $a_{s1010} = -\frac{7}{1000000}$, $a_{s1111} = -\frac{7}{1000000}$, $a_{s19} = -\frac{1}{10}$, $a_{s29} = -1$, $a_{s39} = -\frac{1}{10}$, $a_{s49} = -\frac{1}{10000}$, $a_{s59} = -\frac{1}{10000}$, $a_{s69} = -\frac{1}{10000}$, $a_{s79} = -\frac{1}{10000}$, $a_{s89} = -\frac{1}{10000}$, $a_{s109} = -\frac{1}{100}$, $a_{s119} = -\frac{1}{10}$; the other terms of A_s have a value of 0; (3) we obtain the eigenvalues of $A_{ss} = sI - A_s$ as -7.0×10^{-6}, -7.0×10^{-6}, -7.0×10^{-6}, -7.0×10^{-6}, -7.0×10^{-6}, -7.0×10^{-6}, -7.0×10^{-6}, -7.0×10^{-6}, -7.0×10^{-6}, -7.0×10^{-6}, -0.60001; if the real parts of all the eigenvalues of A_{ss} are negative, then the gain $K \in \Re^{12}$ is correctly chosen, and we can go to step 4; otherwise, we must return to step 1, (4) we choose $P_s = [p_{ij}] \in \Re^{11\times11}$, $p_{11} = 1$, $p_{22} = 71429$, $p_{33} = 1$, $p_{44} = 1$, $p_{55} = 1$, $p_{66} = 1$, $p_{77} = 1$, $p_{88} = 1$, $p_{99} = 1 \times 10^8$, $p_{1010} = 1$, $p_{1111} = 1$, $p_{19} = p_{91} = -0.125$, $p_{29} = p_{92} = -375$, $p_{39} = p_{93} = -0.125$, $p_{49} = p_{94} = -8.7499 \times 10^{-2}$, $p_{59} = p_{95} = -0.62499$, $p_{69} = p_{96} = -8.7499 \times 10^{-2}$, $p_{79} = p_{97} = -0.5$, $p_{89} = p_{98} = -0.125$, $p_{109} = p_{910} = -0.125$, $p_{119} = p_{911} = -0.125$; the other terms of P_s have a value of 0; (5) we substitute $P_s = [p_{ij}] \in \Re^{11\times11}$ and $A_s = A - KC$ into the Lyapunov equation $A_s^T P_s + P_s A_s = -Q_s$, and we find $Q_s = [q_{ij}] \in \Re^{11\times11}$, $q_{11} = 1.4 \times 10^{-5}$, $q_{22} = 1.4 \times 10^{-5}$, $q_{33} = 1.4 \times 10^{-5}$, $q_{44} = 1.4 \times 10^{-5}$, $q_{55} = 1.4 \times 10^{-5}$, $q_{66} = 1.4 \times 10^{-5}$, $q_{77} = 1.4 \times 10^{-5}$, $q_{88} = 1.4 \times 10^{-5}$, $q_{99} = 1.5978 \times 10^8$, $q_{1010} = 1.4 \times 10^{-5}$, $q_{1111} = 1.4 \times 10^{-5}$, $q_{19} = q_{91} = -8.7499 \times 10^{-7}$, $q_{29} = q_{92} = -2.6250 \times 10^{-3}$, $q_{39} = q_{93} = -8.7499 \times 10^{-7}$, $q_{49} = q_{94} = -6.1249 \times 10^{-7}$, $q_{59} = q_{95} = -4.3750 \times 10^{-6}$, $q_{69} = q_{96} = 0.63$, $q_{79} = q_{97} = -3.5000 \times 10^{-6}$, $q_{89} = q_{98} = -8.7499 \times 10^{-7}$, $q_{109} = q_{910} = -8.7499 \times 10^{-7}$, $q_{119} = q_{911} = -8.7499 \times 10^{-7}$; the other terms of Q_s have a value of 0; if $P_s = [p_{ij}] \in \Re^{11\times11}$ and $Q_s = [q_{ij}] \in \Re^{11\times11}$ are positive definite, then $P_s = [p_{ij}] \in \Re^{11\times11}$ is correctly chosen, Theorem 6.1 is met, and we can go to step 6; otherwise, we must return to step 1, 6) we choose the gains $E = [1 \times 10^{-10}, 1 \times 10^{-10}, 1 \times 10^{-10}, 1 \times 10^{-10}, 1 \times 10^{-10}, 1 \times 10^{-10}, 1 \times 10^{-10}, 1 \times 10^{-10}, 1 \times 10^{-10}, 1 \times 10^{-10}, 1 \times 10^{-10}]^T$, $M = [1, 1, 1, 1, 1, 1, 1, 1, 1, 1, 1]^T$, and $\overline{\chi} = 1.73 \times 10^{-5}$ to solve the Estimator 1 as $\dot{\hat{\chi}} = A\hat{\chi} + f(\hat{\chi}, \nu) + K\tilde{y} + E\text{sat}(M\tilde{y})$, $\hat{y} = C\hat{\chi}$, $\hat{u} \cong B^{-1}\left[K\tilde{y} + E\text{sat}(M\tilde{y}) + \overline{\chi}\right]$; if the Estimator 1 reaches an acceptable exactness for the variable and perturbation estimation in the nonlinear process $\dot{\chi} = A\chi + f(\chi, \nu) + Bu$, $y = C\chi$; then the algorithm finishes; otherwise, we must return to step 6.

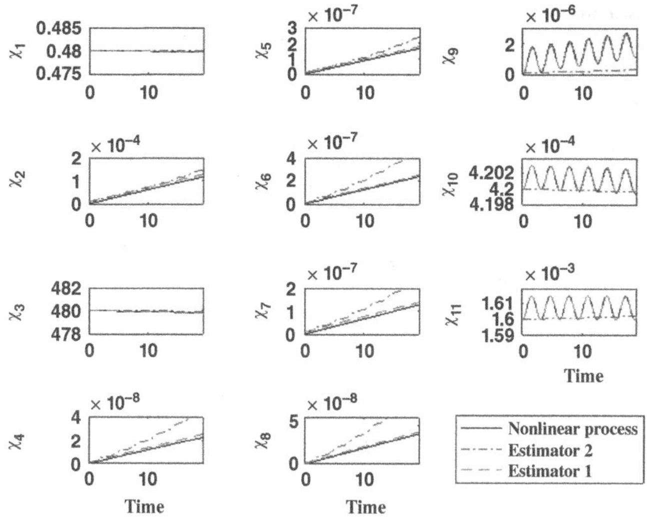

Figure 6.7 The variables estimation in the gasification process.

Figures 6.7 and 6.8 show the variable and perturbation estimation of the Estimator 1 and Estimator 2 applied to the gasification process. Table 6.4 shows the MSE of (6.9).

From Figures 6.7 and 6.8, since the Estimator 1 reaches better variable and perturbation estimation of the process than the Estimator 2, it is seen that the Estimator 1 reaches better performance. In addition, Figures 6.7 and 6.8 show that in the variable and perturbation estimation of Estimator 1, the unwanted chattering is not presented. In Table 6.4, since the MSE is smaller for the Estimator 1 than for the Estimator 2, it is seen that the Estimator 1 reaches better exactness for the variable and perturbation estimation.

Remark 6.5: In Sections 6.3 and 6.4, it would be very tedious and expensive to have the sensors to measure all process variables. So, it highlights one of the major benefits of this chapter; it is that, with the outlet measurements, our estimator can roughly estimate the variable and perturbations.

6.5 Concluding Remarks

In this chapter, an estimator was recommended for the variable and perturbation estimation in nonlinear processes. The variable error convergence was analyzed with the Lyapunov approach. Our estimator was compared to a previous estimator in the gas turbine and gasification processes, concluding that our estimator reached a better performance than the previous estimator for the variable and perturbation estimation. In addition, in our estimator, the unwanted chattering is not

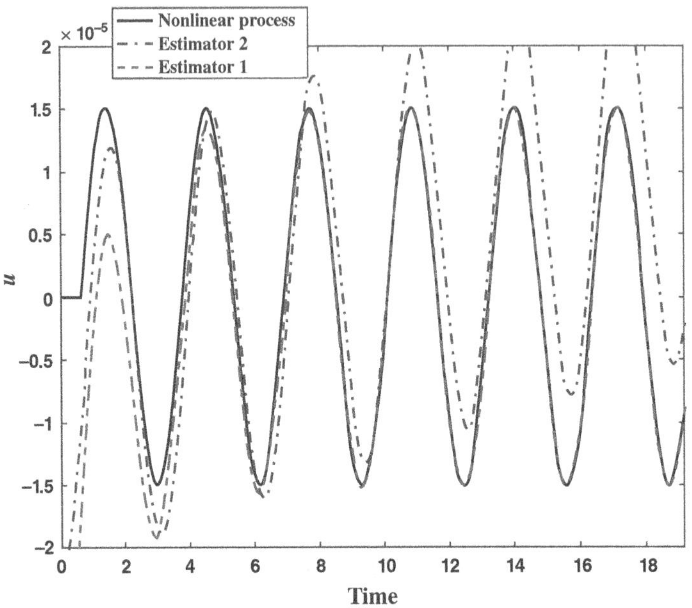

Figure 6.8 The perturbations estimation in the gasification process.

Table 6.4 The gasification process comparisons.

	Estimator 2	Estimator 1
J_{χ}	0.0529	7.5237×10^{-6}
J_y	9.5523×10^{-7}	6.4086×10^{-8}
J_u	1.1859×10^{-5}	1.0118×10^{-5}

presented. Our estimator can be applied to many types of nonlinear processes such as electric, mechanical, hydraulic, or thermal. In the future work, we will seek to use another alternative function that allows us to reduce the unwanted chattering for these type of estimators. We will explore other types of strategies for perturbation estimation, perturbation attenuation, or trajectories reaching in perturbed processes.

References

1 Aguilar Ibañez C., Sira Ramirez H., Acosta J. A. (2017), Stability of active disturbance rejection control for uncertain systems: a Lyapunov perspective, International Journal of Robust and Nonlinear Control, 27, 4541–4553.

2 Aguilar-Ibanez C., Suarez-Castanon M. S. (2019), A trajectory planning based controller to regulate an uncertain 3D overhead crane system, International Journal of Applied Mathematics and Computer Science, 29, 693–702.

3 Pan Y., Li X., Wang H., Yu H. (2018), Continuous sliding mode control of compliant robot arms: a singularly perturbed approach, Mechatronics, 52, 127–134.

4 Pan Y., Yu H. (2015), Dynamic surface control via singular perturbation analysis, Automatica, 57, 29–33.

5 Escobedo Alva J. O., Garcia Estrada E. C., Paramo Carranza L. A., Meda Campaña J. A., Tapia Herrera R. (2018), Theoretical application of a hybrid observer on altitude tracking of quadrotor losing GPS signal, IEEE Access, 6, 76900–76908.

6 Meda Campaña J. A. (2018), On the estimation and control of nonlinear systems with parametric uncertainties and noisy outputs, IEEE Access, 6, 31968–31973.

7 Rubio J. J., Ochoa G., Balcazar R., Pacheco J. (2015), Uniform stable observer for the disturbance estimation in two renewable energy systems, ISA Transactions, 58, 155–164.

8 Rubio J. J. (2016), Hybrid controller with observer for the estimation and rejection of disturbances, ISA Transactions, 65, 445–455.

9 Esfahani P. S., Pieper J. K. (2018), H model predictive control for constrained discrete-time piecewise affine systems, International Journal of Robust and Nonlinear Control, 28, 1973–1995.

10 Esfahani P. S., Pieper J. K. (2019), Robust model predictive control for switched linear systems, ISA Transactions, 89, 1–11.

11 Liu J., Fu M., Xu Y. (2019), Robust synchronization of multiple marine vessels with time-variant disturbance and communication delays, IEEE Access, 7, 39680–39689.

12 Yang P., Ma X., Wang J., Zhang G., Zhang Y., Chen L. (2019), Disturbance observer-based terminal sliding mode control of a 5-DOF upper-limb exoskeleton robot, IEEE Access, 7, 62833–62839.

13 Zhang M., Zhang Y., Cheng X. (2019), Finite-time trajectory tracking control for overhead crane systems subject to unknown disturbances, IEEE Access, 7, 55974–55982.

14 Ji W., Qiu J., Karimi H. R. (2020), Fuzzy-model-based output feedback sliding mode control for discrete-time uncertain nonlinear systems, IEEE Transactions on Fuzzy Systems, 28 (8), 1519–1530.

15 Ji W., Qiu J., Wu L., Lam H. K. (2021), Fuzzy-affine-model-based output feedback dynamic sliding mode controller design of nonlinear systems, IEEE Transactions on Systems, Man and Cybernetics: Systems, 51 (3), 1652–1661.

16 Ferdaus M. M., Pratama M., Anavatti S. G., Garratt M. A., Lughofer E. (2020), Pac: a novel self-adaptive neuro-fuzzy controller for micro aerial vehicles, Information Sciences, 512, 481–505.

17 Liu X., Liu D. (2019), Composite control of nonlinear robotic system with exogenous disturbance, IEEE Access, 7, 19564–19571.

18 Tang Y., Xue Z., Liu X., Han Q., Tuo X. (2019), Leader-following consensus control for multiple fixed-wing UAVs' attitude system with time delays and external disturbances, IEEE Access, 7, 169773–169781.

19 Aguilar Ibañez C., Sossa Azuela J. H., Suarez Castanon M. S. (2016), A backstepping-based procedure with saturation functions to control the pvtol system, Nonlinear Dynamics, 83, 1247–1257.

20 Esfahani P. S., Pieper J. (2020), Machine learning-based piecewise affine model of wind turbines during maximum power point tracking, Wind Energy, 23, 404–422.

21 Rubio J. J., Ochoa G., Mujica Vargas D., Garcia E., Balcazar R., Elias I., Cruz D. R., Juarez C. F., Aguilar A., Novoa J. F. (2019), Structure regulator for the perturbations attenuation in a quadrotor, IEEE Access, 7, 138244–138252.

22 Rubio J. J. (2016), Structure control for the disturbance rejection in two electromechanical processes, Journal of the Franklin Institute, 353, 3610–3631.

23 Tsoutsanis E., Meskin N., Benammar M., Khorasani K. (2014), A component map tuning method for performance prediction and diagnostics of gas turbine compressors, Applied Energy, 135, 572–585.

24 Uppal A. A., Butt S. S., Khan Q., Aschemann H. (2019), Robust tracking of the heating value in an underground coal gasification process using dynamic integral sliding mode control and again-scheduled modified Utkin observer, Journal of Process Control, 73, 113–122.

25 Rubio J. J., Martinez D. I., Garcia V., Gutierrez G. J., Vargas T. M., Ochoa G., Balcazar R., Pacheco J., Meda Campaña J. A., Mujica Vargas D. (2020), The perturbation estimation in two gas plants, IEEE Access, 8 (1), 83081–83091.

7

Regulation of Two Electricity Generators with Dead Zone

7.1 Introduction

The term alternative energy is utilized in the electricity generation from the environment by the utilization of renewable fuels, evading the necessity of a nonrenewable fuel. Some of the most relevant alternative energies can be classified as electricity generators with dead zone based on wind turbines, magnets, or solar panels. The electricity generators with dead zone of this chapter use the interactions between static and dynamic magnets for the electricity generation [1, 2]. The motivation of this chapter is the regulation of electricity generators with dead zone. A regulator is one strategy for the regulation of electricity generators with dead zone, which lets us to reach some constant behaviors for the angle position, angle speed, and current [3–8].

There are many kinds of nonlinearities which impede to reach the regulation of electricity generators with dead zone. Some examples of these nonlinearities are the arbitrary switching [9–13], the time-delays [14–18], the impulse perturbations [19, 20], or the unknown nonlinearities [21–24]. The major issue is that in most of the cases, the mentioned nonlinearities are unknown. The challenge of this chapter is the regulation of electricity generators with dead zone being affected by unknown nonlinearities.

The magnets interaction nonlinearities and generators nonlinearities are two unknown nonlinearities of the electricity generators with dead zone, which can severely affect their performance [21–24]. So, the research about regulators for the regulation of electricity generators with dead zone, where the magnets interaction nonlinearities and generators nonlinearities are unknown, is of interest. Two alternatives are the proportional-derivative (PD) regulator of [3–5] and the sliding-mode (SM) regulator of [6–8].

In this chapter, an SM regulator with sine function (SMWS) is recommended for the regulation of electricity generators with dead zone, being affected by magnets interaction nonlinearities and generators nonlinearities. To reach this goal, our recommended regulator has the following benefits:

1. The SM regulator starts from the proportional derivative regulator with one additional term of the signum function utilized to reach the regulation. The SM regulator with sine function starts from the SM regulator with the modifications that the saturation function is utilized to reach a smoother performance instead of the signum function, and the sine function is applied to reach an upper bound in the proportional gain error.
2. Since the SM strategy is utilized in the recommended regulator instead of the magnets interaction nonlinearities and generators nonlinearities, the knowledge of these nonlinearities is not required in the electricity generators with dead zone. Our recommended regulator is also utilized to reach some chosen constant behaviors for the angle position, angle speed, and current.
3. We warrant the regulation of the regulator error in our recommended regulator based on the Lyapunov approach. This chapter is concentrated on the asymptotic regulation.

We organize the chapter as follows. In Section 7.2, we present the mathematical model of electricity generators with dead zone. In Section 7.3, we present the SM regulator with sine function for the regulation of electricity generators with dead zones. In Section 7.4, we simulate the recommended regulator for the regulation of two electricity generators with dead zone. In Section 7.5, we present the conclusions and future research.

7.2 Mathematical Model and Regulators for the Electricity Generators with Dead Zone

The descriptions of the electricity generators with dead-zone symbols are shown in Table 7.1, the parameter values for mathematical models are shown in Table 7.2. The descriptions of the mathematical model symbols are shown in Table 7.3, and the descriptions of the regulators symbols are shown in Table 7.4.

In this section, we address some concepts such as the mathematical model and regulators for the electricity generators with dead zones.

We consider the electricity generators with dead zone with two static magnets and two dynamic magnets as shown in Figure 7.1 via the utilization of two static magnets and two dynamic magnets.

Table 7.1 Description of the electricity generators with dead-zone symbols.

Parameter	Description
m_e	The mass of the link 1
a	The width of the link 1
c	The length of the link 1
m_i	The mass of the dynamic magnet
r_i	The radius of the surface in the magnet
m	The mass
r_g	The radius of the body rotation
B_{11}	The magnet flux density produced by dynamic magnet 1 and static magnet 1
B_{12}	The magnet flux density produced by dynamic magnet 1 and static magnet 2
B_{21}	The magnet flux density produced by dynamic magnet 2 and static magnet 1
B_{22}	The magnet flux density produced by dynamic magnet 2 and static magnet 2
μ	The magnet permeability constant in magnets
h	The distance of a dynamic magnet and a static magnet on the axis z
K_m	The electromotive force constant
R	The resistance
i	The current
L	The inductance
R_e	The load resistance

Table 7.2 Parameters of electricity generators with dead zone.

Parameter	Value	Parameter	Value
m_i	9.2×10^{-2} kg	R_e	30 Ω
μ	9.42×10^{-5} H/m	r_g	7.5×10^{-2} m
B	8.4×10^{-1} T	r_i	2.9×10^{-2} m
L	6.03×10^{-1} H	a	1.5×10^{-1} m
K_m	45×10^{-2} V s/rad	c	1.5×10^{-2} m
b	1×10^{-1} kg m^2 rad/s	n_r	0.5
m_e	5×10^{-3} kg	n_l	0.5
h	4×10^{-2} m	w_r	0.5
R	6.96 Ω	w_l	−0.5

Table 7.3 Description of the mathematical model symbols.

Symbol	Description
$\chi_1 \in \Re$	The angle position
$\chi_2 \in \Re$	The angle speed
$\chi_3 \in \Re$	The current
$\nu \in \Re$	The fictitious input of torque produced by the magnets interaction
$O(\chi_1) \in \Re$	The magnets interaction nonlinearities
χ_1, χ_2, χ_3	The states
Q, J, X, Y	Scalar constants
$u \in \Re$	The input of the generator perturbations
n_r, n_l, w_r, w_l	Constant terms for the generator perturbations
$P(u)$	The generator perturbations

Table 7.4 Description of the regulators symbols.

Symbol	Description
$\widetilde{\chi}_1 = \chi_1 - \chi_1^d \in \Re$	Thc position regulator error
$\chi_1^d \in \Re$	The constant reference position
$\widetilde{\chi}_2 = \chi_2 \in \Re$	The speed regulator error
$K_1 \in \Re$	The proportional gain
$K_2 \in \Re$	The derivative gain
$\mathrm{sign}(\cdot)$	The signum function
$\mathrm{sat}(\cdot)$	The saturation function
$\sin(\cdot)$	The sine function
$\cos(\cdot)$	The cosine function
K	The SM gain
$\widetilde{\chi}_1^2 = (\chi_1 - \chi_1^d)^2$	The position regulator error
$\widetilde{\chi}_2^2 = \chi_2^2$	The speed regulator error
$\widetilde{\chi}_3^2 = \chi_3^2$	The current regulator error

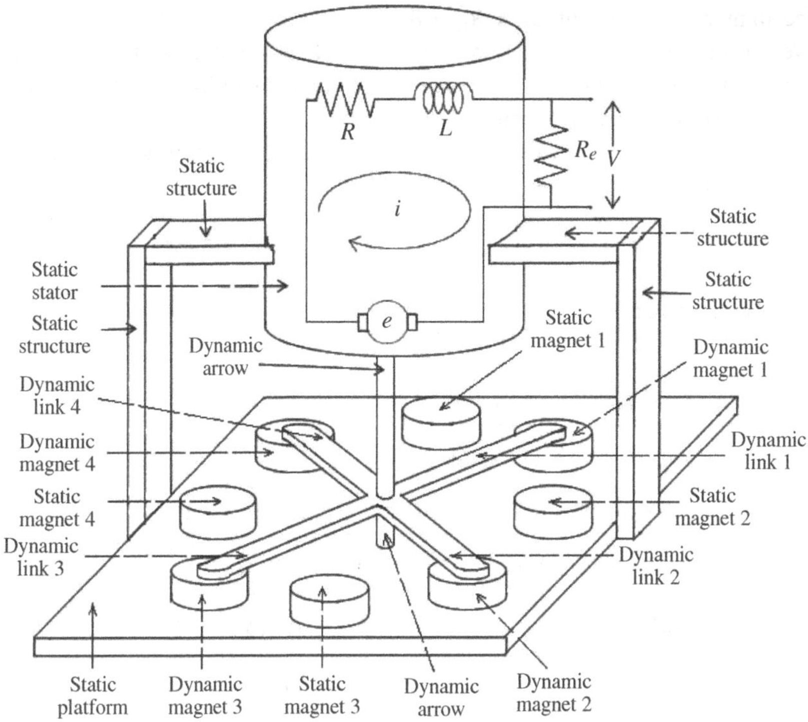

Figure 7.1 Electricity generators with dead zone with static magnets and dynamic magnets.

We express the electricity generators with dead zone with two static magnets and two dynamic magnets as [1, 2]:

$$\dot{\chi}_1 = \chi_2,$$

$$\dot{\chi}_2 = \left(\frac{1}{m_e\left(\frac{1}{24}a^2 + \frac{1}{24}c^2 + r_g^2\right) + m_i\left(r_i^2 + 2r_g^2\right)} \right)$$

$$* \left(\nu - \frac{\pi B^2 r_i^2 r_g^2 \sin \chi_1}{\mu\left[h^2 + 2r_g^2(1 - \cos \chi_1)\right]^{0.5}} \right. \tag{7.1}$$

$$\left. + \frac{\pi B^2 r_i^2 r_g^2 \sin \chi_1}{\mu\left[h^2 + 2r_g^2(1 + \cos \chi_1)\right]^{0.5}} - b\chi_2 \right),$$

$$\dot{\chi}_3 = -\frac{(R + R_e)}{L}\chi_3 + \frac{K_m}{L}\chi_2,$$

since all magnets are equal, $B_{11} = B_{12} = B$.

We consider the electricity generators with dead zone with four static magnets and four dynamic magnets as shown in Figure 7.1 via the utilization of four static magnets and four dynamic magnets.

We express the electricity generators with dead zone with four static magnets and four dynamic magnets as [1, 2]:

$$
\begin{aligned}
\dot{\chi}_1 &= \chi_2,\\
\dot{\chi}_2 &= \left(\frac{1}{m_e\left(\frac{1}{24}a^2 + \frac{1}{24}c^2 + r_g^2\right) + m_i\left(r_i^2 + 2r_g^2\right)}\right)\\
&* \left(\nu - \frac{\pi B^2 r_i^2 r_g^2 \sin\chi_1}{\mu\left[h^2 + 2r_g^2(1 - \cos\chi_1)\right]^{0.5}}\right.\\
&- \frac{\pi B^2 r_i^2 r_g^2 \cos\chi_1}{\mu\left[h^2 + 2r_g^2(1 + \sin\chi_1)\right]^{0.5}} + \frac{\pi B^2 r_i^2 r_g^2 \sin\chi_1}{\mu\left[h^2 + 2r_g^2(1 + \cos\chi_1)\right]^{0.5}}\\
&+ \frac{\pi B^2 r_i^2 r_g^2 \cos\chi_1}{\mu\left[h^2 + 2r_g^2(1 - \sin\chi_1)\right]^{0.5}} - \frac{\pi B^2 r_i^2 r_g^2 \sin\left(\chi_1 - \frac{\pi}{4}\right)}{\mu\left[h^2 + 2r_g^2\left(1 - \cos\left(\chi_1 - \frac{\pi}{4}\right)\right)\right]^{0.5}}\\
&- \frac{\pi B^2 r_i^2 r_g^2 \cos\left(\chi_1 - \frac{\pi}{4}\right)}{\mu\left[h^2 + 2r_g^2\left(1 + \sin\left(\chi_1 - \frac{\pi}{4}\right)\right)\right]^{0.5}} + \frac{\pi B^2 r_i^2 r_g^2 \sin\left(\chi_1 - \frac{\pi}{4}\right)}{\mu\left[h^2 + 2r_g^2\left(1 + \cos\left(\chi_1 - \frac{\pi}{4}\right)\right)\right]^{0.5}}\\
&\left.+ \frac{\pi B^2 r_i^2 r_g^2 \cos\left(\chi_1 - \frac{\pi}{4}\right)}{\mu\left[h^2 + 2r_g^2\left(1 - \sin\left(\chi_1 - \frac{\pi}{4}\right)\right)\right]^{0.5}} - b\chi_2\right),\\
\dot{\chi}_3 &= -\frac{(R + R_e)}{L}\chi_3 + \frac{K_m}{L}\chi_2,
\end{aligned}
\tag{7.2}
$$

since all magnets are equal, $B_{11} = B_{12} = B_{21} = B_{22} = B$.

From the electricity generators with dead zone with two static magnets and two dynamic magnets (7.1) and the electricity generators with dead zone with four static magnets and four dynamic magnets (7.2), we express the mathematical model for the electricity generators with dead zone as:

$$\begin{aligned}&\dot{\chi}_1 = \chi_2,\\&Q\dot{\chi}_2 + J\chi_2 + O(\chi_1) = \nu,\\&\dot{\chi}_3 + X\chi_3 + Y\chi_1 = 0.\end{aligned} \tag{7.3}$$

The fictitious input ν is affected by the generator perturbations. It is like a chattering movement presented in the electricity generators with dead zone. We express the fictitious input ν as [6]:

$$\nu = \begin{cases} n_r(u - w_r) & u \geq w_r \\ 0 & w_l < u < w_r . \\ n_l(u - w_l) & u \leq w_l \end{cases} \tag{7.4}$$

We see that the model (7.1) can be expressed as the model (7.3) and (7.4) with:

$$\begin{aligned}&Q = m_e\left(\frac{1}{24}a^2 + \frac{1}{24}c^2 + r_g^2\right) + m_i(r_i^2 + 2r_g^2),\\&J = b,\\&X = \frac{(R + R_e)}{L},\\&Y = -\frac{K_m}{L},\\&O(\chi_1) = \frac{\pi B^2 r_i^2 r_g^2 \sin\chi_1}{\mu\left[h^2 + 2r_g^2(1 - \cos\chi_1)\right]^{0.5}} - \frac{\pi B^2 r_i^2 r_g^2 \sin\chi_1}{\mu\left[h^2 + 2r_g^2(1 + \cos\chi_1)\right]^{0.5}}.\end{aligned} \tag{7.5}$$

We see that the model (7.2) can be expressed as the model (7.3) and (7.4) with:

$$\begin{aligned}&Q = m_e\left(\frac{1}{24}a^2 + \frac{1}{24}c^2 + r_g^2\right) + m_i(r_i^2 + 2r_g^2),\\&J = b,\\&X = \frac{(R + R_e)}{L},\\&Y = -\frac{K_m}{L},\end{aligned}$$

$$O(\chi_1) = \left(\frac{\pi B^2 r_i^2 r_g^2 \sin \chi_1}{\mu\left[h^2 + 2r_g^2(1 - \cos \chi_1)\right]^{0.5}} + \frac{\pi B^2 r_i^2 r_g^2 \cos \chi_1}{\mu\left[h^2 + 2r_g^2(1 + \sin \chi_1)\right]^{0.5}} \right.$$

$$- \frac{\pi B^2 r_i^2 r_g^2 \sin \chi_1}{\mu\left[h^2 + 2r_g^2(1 + \cos \chi_1)\right]^{0.5}} - \frac{\pi B^2 r_i^2 r_g^2 \cos \chi_1}{\mu\left[h^2 + 2r_g^2(1 - \sin \chi_1)\right]^{0.5}}$$

$$+ \frac{\pi B^2 r_i^2 r_g^2 \sin\left(\chi_1 - \frac{\pi}{4}\right)}{\mu\left[h^2 + 2r_g^2\left(1 - \cos\left(\chi_1 - \frac{\pi}{4}\right)\right)\right]^{0.5}} + \frac{\pi B^2 r_i^2 r_g^2 \cos\left(\chi_1 - \frac{\pi}{4}\right)}{\mu\left[h^2 + 2r_g^2\left(1 + \sin\left(\chi_1 - \frac{\pi}{4}\right)\right)\right]^{0.5}}$$

$$\left. - \frac{\pi B^2 r_i^2 r_g^2 \sin\left(\chi_1 - \frac{\pi}{4}\right)}{\mu\left[h^2 + 2r_g^2\left(1 + \cos\left(\chi_1 - \frac{\pi}{4}\right)\right)\right]^{0.5}} - \frac{\pi B^2 r_i^2 r_g^2 \cos\left(\chi_1 - \frac{\pi}{4}\right)}{\mu\left[h^2 + 2r_g^2\left(1 - \sin\left(\chi_1 - \frac{\pi}{4}\right)\right)\right]^{0.5}} \right). \tag{7.6}$$

The generator perturbations of electricity generators with dead zone are symmetric as $n_r = n_l$ in (7.4); so, we can express the fictitious input ν (7.4) as:

$$\nu = \begin{cases} n_l u & u \geq w_r \\ n_l u & w_l < u < w_r \\ n_l u & u \leq w_l \end{cases} + \begin{cases} -n_l w_r & u \geq w_r \\ -n_l u & w_l < u < w_r. \\ -n_l w_l & u \leq w_l \end{cases} \tag{7.7}$$

After some mathematical operations, the fictitious input ν (7.7) is expressed as follows:

$$\nu = n_l u - P(u),$$

$$P(u) = \begin{cases} n_l w_r & u \geq w_r \\ n_l u & w_l < u < w_r. \\ n_l w_l & u \leq w_l \end{cases} \tag{7.8}$$

We notice that the generator perturbations $P(u)$ are bounded as follows:

$$|P(u)| \leq \overline{P}. \tag{7.9}$$

Remark 7.1: Even the mathematical model for the electricity generators with dead zone (7.3) and (7.4) seems to be simple. The complexity in this mathematical model is mainly concentrated on the magnets interaction nonlinearities $O(\chi_1) \in \Re$, such that the design and validation of this mathematical model is in two publications [1, 2].

Remark 7.2: From (7.5) and (7.6), it is observed that the magnets interaction nonlinearities $O(\chi_1) \in \Re$ are the most complex part in the mathematical model (7.3) and (7.4), while Q, J, X, and Y are scalar constants corresponding to the electricity generators with dead zone. So, these magnet interaction nonlinearities $O(\chi_1) \in \Re$ are mainly concentrated in this chapter.

Now, we express the proportional derivative and SM regulators because they will be utilized for the comparisons in the Section 7.4.

We express the proportional derivative regulator as [3–5]:

$$u = \frac{1}{n_l}\{-K_1\tilde{\chi}_1 - K_2\tilde{\chi}_2\}. \tag{7.10}$$

We express the SM regulator as [6–8]:

$$u = \frac{1}{n_l}\{-K_1\tilde{\chi}_1 - K_2\tilde{\chi}_2 - K\text{sign}(\tilde{\chi}_2)\},$$
$$\text{sign}(\tilde{\chi}_2) = \begin{cases} 1 & \tilde{\chi}_2 > 0 \\ 0 & \tilde{\chi}_2 = 0. \\ -1 & \tilde{\chi}_2 < 0 \end{cases} \tag{7.11}$$

7.3 Sliding-mode Regulator with Sine Function

In this section, we address the design and regulation of the SM regulator with sine function.

We notice that the magnet interaction nonlinearities $O(\chi_1)$ (7.3) are bounded as:

$$|O(\chi_1)| \le \overline{O}. \tag{7.12}$$

Since we use the reference speed states as $\chi_2^d = 0$, we consider the regulation case and the $\chi_1^d \in \Re$ reference position as a constant in this chapter. We express the SM regulator with sine function u as:

$$u = \frac{1}{n_l}\{-\cos(\tilde{\chi}_1)^T K_1 \sin(\tilde{\chi}_1) - K_2\tilde{\chi}_2 - K\text{sat}(\tilde{\chi}_2)\},$$
$$\text{sat}(\tilde{\chi}_2) = \begin{cases} 1 & \tilde{\chi}_2 > 1 \\ \tilde{\chi}_2 & |\tilde{\chi}_2| \le 1. \\ -1 & \tilde{\chi}_2 < -1 \end{cases} \tag{7.13}$$

K is a constant such as $\overline{O} + \overline{P} \le K$, $\overline{O}$ is in (7.12) and $\overline{P}$ is in (7.9). It is relevant to note that we do not know the behaviors of $O(\chi_1)$ and $P(u)$, and we use their upper bounds $\overline{O}$ and $\overline{P}$.

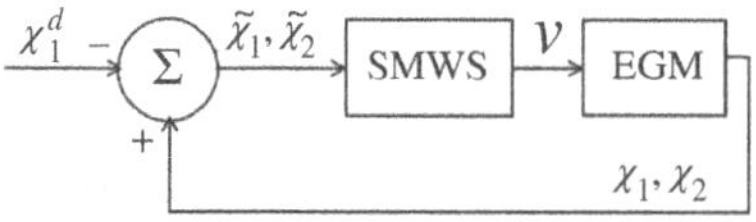

Figure 7.2 SM regulator with sine function.

Remark 7.3: As we consider the regulation case, we do not require the application of the sine function to the term $K_2\widetilde{\chi}_2$, i.e., since $\chi_2^d = 0$ and $\widetilde{\chi}_2 = \chi_2$, we obtain small values in $\widetilde{\chi}_2$. Since the state χ_3 is dependent of the other states χ_1, χ_2, the regulation of the states χ_1, χ_2 yields the regulation of the state χ_3, and we do not require the state χ_3 in the regulator.

Remark 7.4: The convergence speed of the recommended regulator can be affected by choosing the gains K_1, K_2, and K. The complexity in the implementation of the recommended regulator is in the programming of the saturation function sat(·); nevertheless, this complexity is also in all the regulators, which use functions.

In Figure 7.2, we show the SM regulator with sine function (7.9), (7.12), and (7.13), named SMWS for the regulation in the electricity generators with dead zone (7.3) and (7.4), named electricity generators with magnets (EGM).

It is shown that the PD regulator (7.10) and [3–5], the SM regulator (7.9), (7.11), and (7.12) and [6–8], and the recommended SM regulator with sine function (7.9), (7.12), and (7.13), use the proportional gain K_1 and derivative gain K_2. In addition, the existence, similarities, and differences between PD, SM, and SMWS are detailed as: the SM regulator starts from the PD regulator with one additional term of the signum function utilized to reach the regulation; and the SM regulator with sine function starts from the SM regulator with the modifications that the saturation function is utilized to reach a smoother performance instead of the signum function; and the sine function is applied to reach an upper bound of the proportional gain error.

Now, we will discuss the regulation of the regulator in Theorem 7.1

Theorem 7.1: *The regulator error in the close-loop model of the SM regulator with sine function (7.9), (7.12), and (7.13), and electricity generators with dead zone (7.3) and (7.4) is stabilized, and the speed regulator error $\widetilde{\chi}_2$ will converge to:*

$$\limsup_{T\to\infty} \| \widetilde{\chi}_2 \|^2 = 0, \tag{7.14}$$

with T as the final time, $\widetilde{\chi}_2 = \chi_2$, the magnets interaction nonlinearities and generator perturbations are bounded as $|O(\chi_1)| \leq \overline{O}$, $|P(u)| \leq \overline{P}$, and $\overline{O} + \overline{P} \leq K$.

Proof. See [25] for the proof. ■

Remark 7.5: From (7.1) and (7.2), it can be seen that the angle speed behavior affects the angle position and current behaviors in the electricity generators with dead zone. So, the angle speed convergence (7.14) is utilized to reach some chosen constant behaviors for the angle position and current in the electricity generators with dead zone.

Remark 7.6: This chapter is concentrated on electricity generators with dead zone. Nevertheless, our recommended regulator could be applied to other plants with similar structure to (7.3) and (7.4), such as the motors, machines, robots, pendulums, or cranes.

Remark 7.7: There are several regulation works, such as the exponential regulation [14, 17, 18, 20], the asymptotic regulation [5, 6, 16, 22, 23], the uniformly ultimately boundedness [9–11], the feedback regulation [15, 24], the stochastic regulation [12, 21], or the mean square regulation [13]. This chapter is concentrated on the asymptotic regulation.

7.4 Simulations

In this section, we compare the SMWS (7.9), (7.12), and (7.13), , the PD regulator (7.9), (7.10), and (7.12), [3–5], and the SM regulator (7.9), (7.11), and (7.12), [6–8], for the regulation of the two electricity generators with dead zone (7.3) and (7.4). Our goal in the regulators is that the angle position, angle speed, and current in electricity generators with dead zone must reach some chosen constant references for the angle position, angle speed, and current as fast as possible in presence of the magnets interaction nonlinearities and generator perturbations. The chosen constant references in the angle position are higher or equal to zero, while the chosen constant references in the angle speed and current are equal to zero. We utilize the root mean square error (RMSE) for the comparisons as follows:

$$\mathrm{RMSE} = \left(\frac{1}{T} \int_{0}^{T} \widetilde{\chi}^2 dt \right)^{\frac{1}{2}}, \tag{7.15}$$

with $\widetilde{\chi}^2 = \widetilde{\chi}_1^2 + \widetilde{\chi}_2^2 + \widetilde{\chi}_3^2$ for the states error or $\widetilde{\chi}^2 = \nu^2$ for the fictitious input error.

7.4.1 Electricity Generators with Dead Zone with Two Static Magnets and Two Dynamic Magnets

We express the electricity generators with dead zone with two static magnets and two dynamic magnets in (7.1), (7.3), (7.4), and (7.5).

PD of [3–5] is expressed by (7.9), (7.10), and (7.12), with parameters $K_1 = 100$ and $K_2 = 0.1$.

SM of [6–8] is expressed by (7.9), (7.11), and (7.12) with parameters $K_1 = 100$, $K_2 = 0.1$ and $K = 0.5$.

SMWS is expressed by (7.9), (7.12), and (7.13), with parameters $K_1 = 100$, $K_2 = 0.1$ and $K = 0.5$.

We compare the fictitious input in Figure 7.3 and a zoom in Figure 7.4, we compare the states in Figure 7.5 and a zoom in Figure 7.6, and we show the RMSE (7.15) in Figure 7.7 and Table 7.5 for the electricity generators with dead zone with two static magnets and two dynamic magnets.

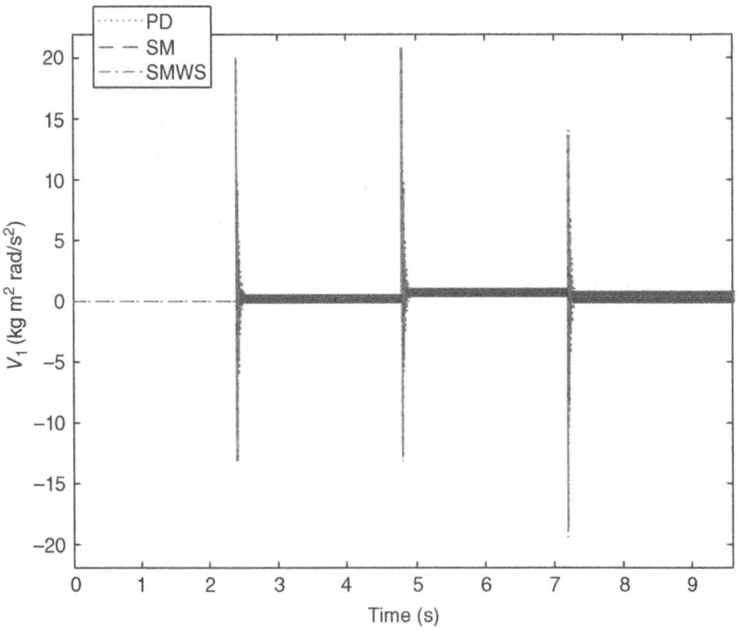

Figure 7.3 Fictitious input of the first electricity generators with dead zone.

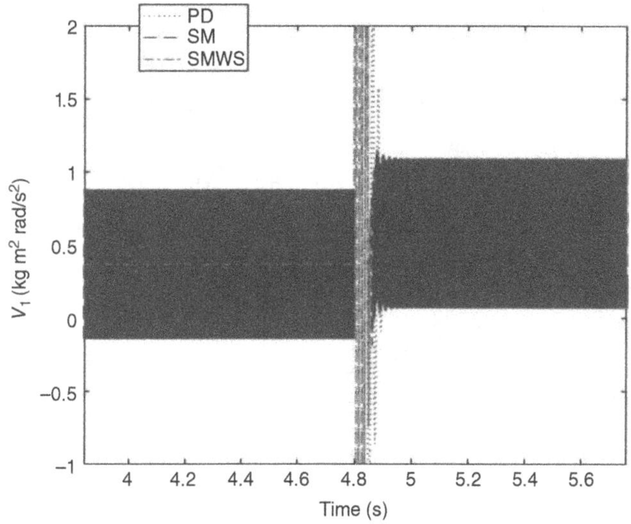

Figure 7.4 A zoom in the fictitious input of the first electricity generators with dead zone.

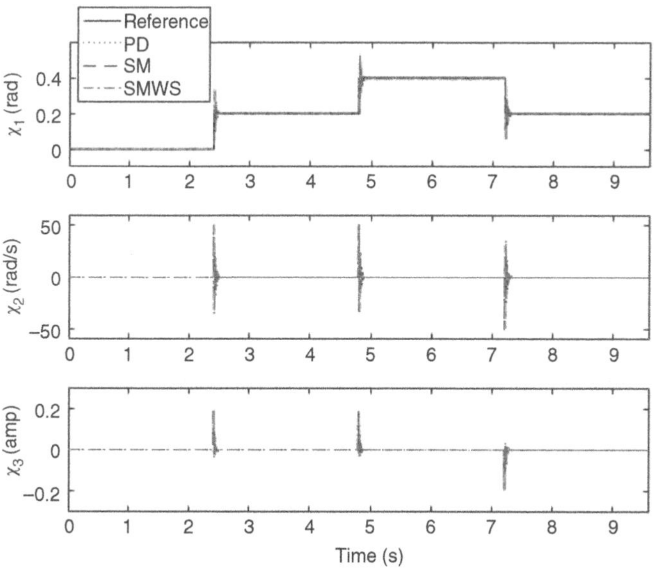

Figure 7.5 States of the first electricity generators with dead zone.

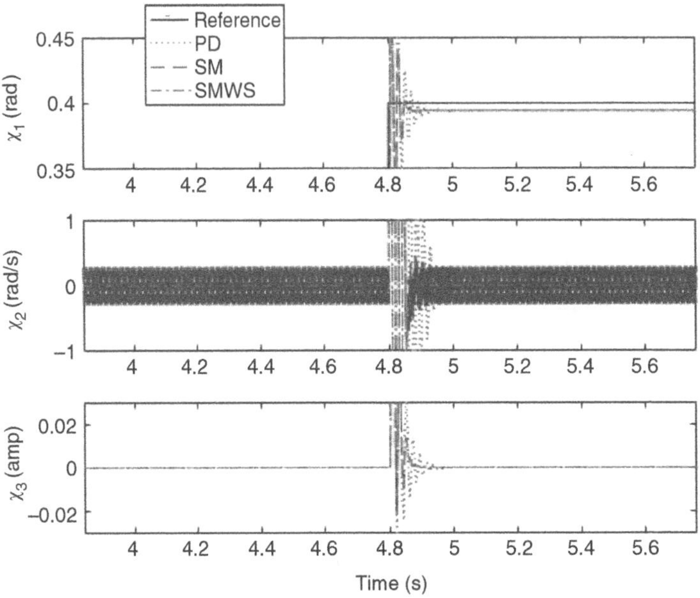

Figure 7.6 A zoom in the states of the first electricity generators with dead zone.

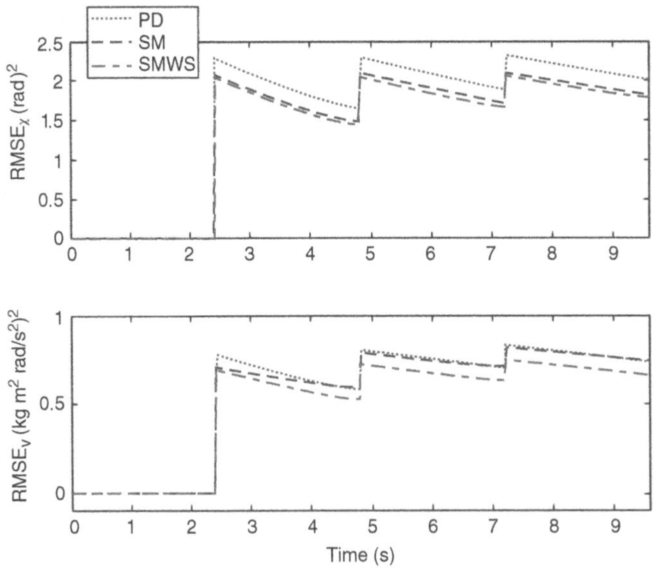

Figure 7.7 RMSE of the first electricity generators with dead zone.

Table 7.5 Comparisons for first electricity generators with dead zone.

Strategies	RMSE for $\tilde{\chi}$	RMSE for ν
PD	2.0154	0.7386
SM	1.8226	0.7463
SMWS	1.7835	0.6656

In Figures 7.5 and 7.6, since the angle position, angle speed, and current of SMWS reach the constant references in the angle position, angle speed, and current better than the angle position, angle speed, and current of PD and SM, we can see that SMWS has the best performance. In Figures 7.3 and 7.4, the fictitious input for SMWS is smaller than for PD and SM, and in Figure 7.7 and Table 7.5, since the RMSE for the SMWS is smaller than for PD and SM, we can show that SMWS has the best performance.

7.4.2 Electricity Generators with Dead Zone with Four Static Magnets and Four Dynamic Magnets

We express the electricity generators with dead zone with four static magnets and four dynamic magnets in (7.2)–(7.4) and (7.6).

PD of [3–5] is expressed by (7.9), (7.10), and (7.12), with parameters $K_1 = 100$, $K_2 = 0.1$.

SM of [6–8] is expressed by (7.9), (7.11), and (7.12), with parameters $K_1 = 100$, $K_2 = 0.1, K = 0.5$.

SMWS is expressed by (7.9), (7.12), and (7.13), with parameters $K_1 = 100$, $K_2 = 0.1, K = 0.5$.

We compare the fictitious inputs in Figure 7.8 and a zoom in Figure 7.9, we compare the states in Figure 7.10 and a zoom in Figure 7.11, and we show the RMSE (7.15) in Figure 7.12 and Table 7.6 for the electricity generators with dead zone with four static magnets and four dynamic magnets.

In Figures 7.10 and 7.11, since the angle position, angle speed, and current of SMWS reach the constant references in the angle position, angle speed, and current better than the angle position, angle speed, and current of PD and SM, we can see that SMWS has the best performance. In Figures 7.8 and 7.9, the fictitious input for SMWS is smaller than for PD and SM, and in Figure 7.12 and Table 7.6, since the RMSE for the SMWS is smaller than for PD and SM, we can show that SMWS has the best performance.

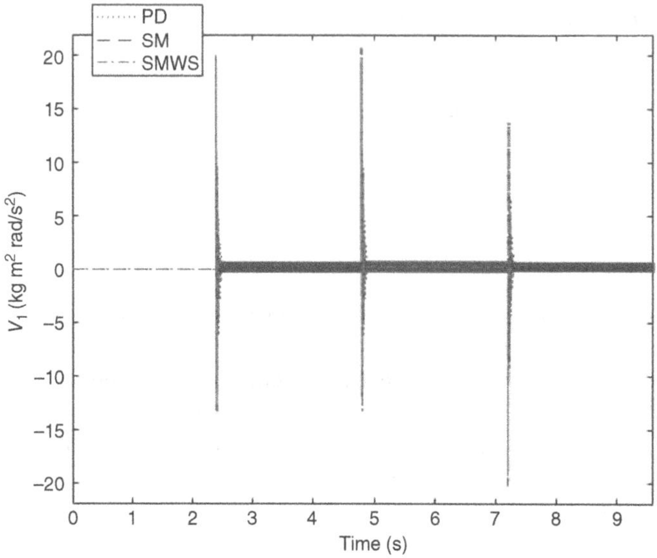

Figure 7.8 Fictitious input of the second electricity generators with dead zone.

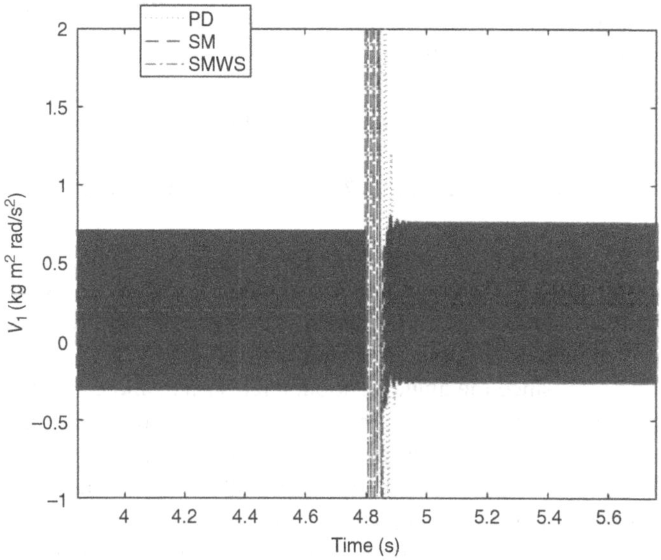

Figure 7.9 A zoom in the fictitious input of the second electricity generators with dead zone.

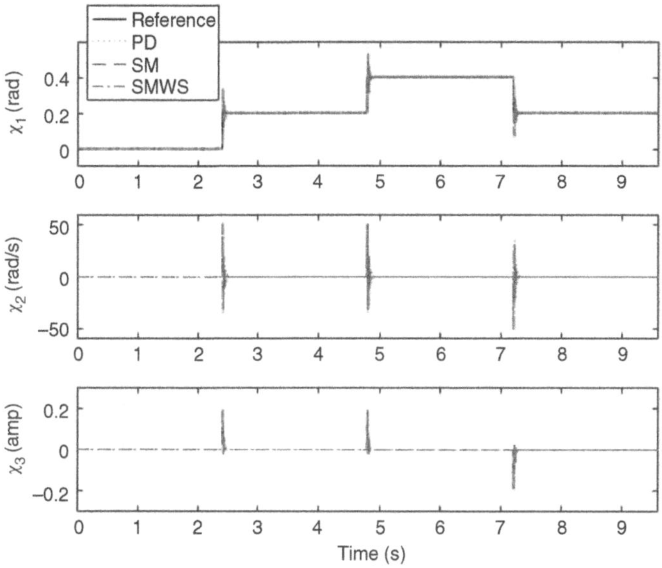

Figure 7.10 States of the second electricity generators with dead zone.

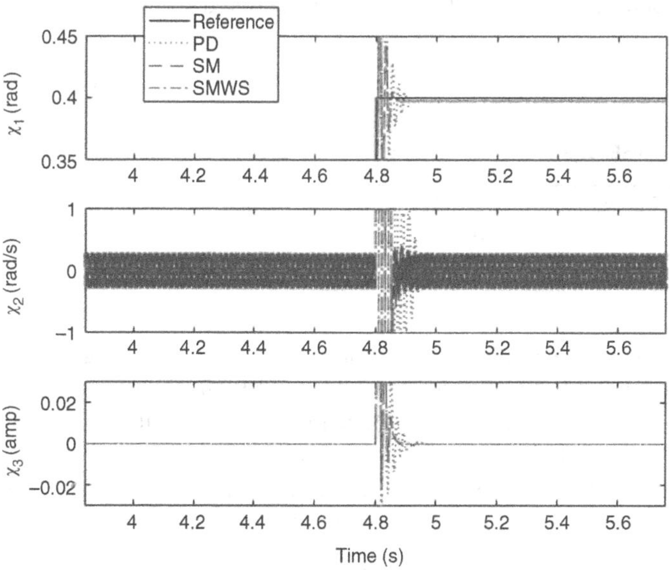

Figure 7.11 A zoom in the states of the second electricity generators with dead zone.

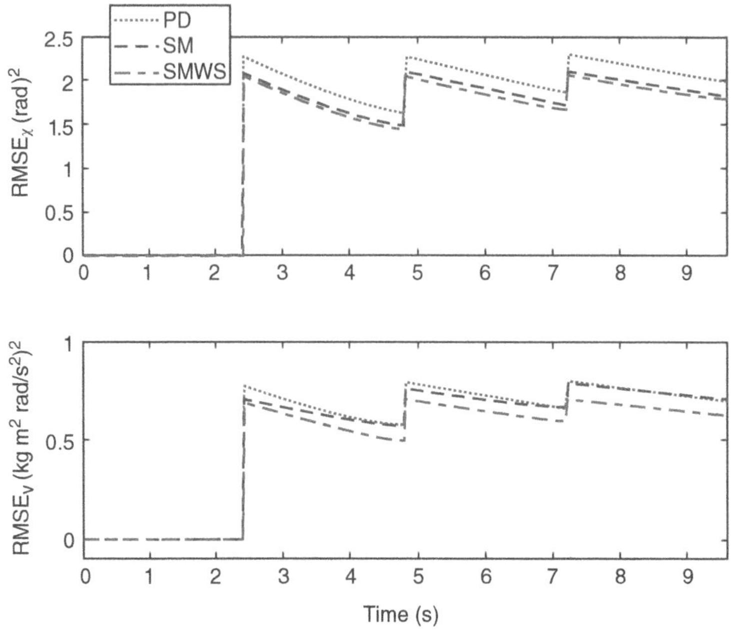

Figure 7.12 RMSE of the second electricity generators with dead zone.

Table 7.6 Comparisons for second electricity generators with dead zone.

Strategies	RMSE for $\widetilde{\chi}$	RMSE for v
PD	2.0108	0.6995
SM	1.8230	0.7098
SMWS	1.7821	0.6224

Remark 7.8: The course of RMSE is not monotonically increasing in Figures 7.7 and 7.12, even though it is indirectly recommended in (7.15) based on the following two reasons: (a) the final time T should increase for more simulation time, it is the denominator of (7.15); (b) our goal is that the states error $\widetilde{\chi}^2 = \widetilde{\chi}_1^2 + \widetilde{\chi}_2^2 + \widetilde{\chi}_3^2$ or the fictitious input error $\widetilde{\chi}^2 = v^2$ should decrease for more simulation time, it is the numerator of (7.15). The mentioned reasons yield the decreasing of RMSE.

Remark 7.9: Since Figures 7.3, 7.4, 7.8, and 7.9 describe the impulses utilized as the excitation to start moving these electricity generators with dead zone, and Figures 7.5, 7.6, 7.10, and 7.11 describe the angle position, angle speed, and current utilized as the responses of these electricity generators with dead zone, the simulations could represent the real experiments.

7.5 Concluding Remarks

In this chapter, we suggest a sliding mode regulator with sine function for the regulation of two electricity generators with dead zone, being affected by unknown nonlinearities. In the simulations, we show that since the angle position, angle speed, and current of our SM regulator with sine function reach the constant references of angle position, angle speed, and current better than the angle position, angle speed, and current of the proportional derivative and SM regulators, we can see that recommended regulator has the best performance in presence of the magnets interaction nonlinearities and generator perturbations. Our regulator illustrates the viability, efficiency, and potential of the approach, especially relevant in electricity generators with dead zone. Our recommended strategy could also be applied to other mechatronic systems such as the electricity motors, machines, robots, pendulums, or cranes. The drawback of our regulator is that it is designed for models with structure of electricity generators with dead zone; for models with other structures, the design should require modifications. In the future, we will apply our recommended regulator for the regulation of robots or pendulums being affected by unknown nonlinearities.

References

1 Rubio J. J., Aguilar A., Meda-Campaña J. A., Ochoa G., Balcazar R., Lopez J. (2019), An electricity generator based on the interaction of static and dynamic magnets, IEEE Transactions on Magnetics, 55, 8204511.

2 Aguilar A., Rubio J. J., Meda-Campaña J. A., Martinez D. I., Vargas T. M., Garcia V., Novoa J. F., Mujica-Vargas D., Pacheco J., Gutierrez G. J. (2020), Movable and immovable magnets for two machines, International Journal of Applied Electromagnetics and Mechanics, 63, 229–248.

3 Escobedo-Alva J. O., Garcia-Estrada E. C., Paramo-Carranza L. A., Meda-Campaña J. A., Tapia-Herrera R. (2018), Theoretical application of a hybrid observer on altitude tracking of quadrotor losing gps signal, IEEE Access, 6, 76900–76908.

4 Meda-Campaña J. A. (2018), On the estimation and control of nonlinear systems with parametric uncertainties and noisy outputs, IEEE Access, 6, 31968–31973.

5 Rubio J. J., Zamudio Z., Pacheco J., Mujica Vargas D. (2013), Proportional derivative control with inverse dead-zone for pendulum systems, Mathematical Problems in Engineering, 2013, 1–9.

6 Rubio J. J. (2017), Sliding mode control of robotic arms with deadzone, IET Control, Theory and Applications, 11, 1214–1221.

7 Aguilar-Ibanez C., Sira-Ramirez H., Suarez-Castanon M. S., Garrido R. (2019), Robust trajectory-tracking control of a pvtol under crosswind, Asian Journal of Control, 21, 1293–1306.

8 Aguilar-Ibanez C., Suarez-Castanon M. S. (2019), A trajectory planning based controller to regulate an uncertain 3D overhead crane system, International Journal of Applied Mathematics and Computer Science, 29, 693–702.
9 Chang Y., Wang Y., Alsaadi F. E., Zong G. (2019), Adaptive fuzzy output-feedback tracking control for switched stochastic pure-feedback nonlinear systems, International Journal of Adaptive and Control Signal Processing, 33, 1567–1582.
10 Chang Y., Zhang S., Alotaibi N. D., Alkhateeb A. F. (2020), Observer-based adaptive finite-time tracking control for a class of switched nonlinear systems with unmodeled dynamics, IEEE Access, 8, 204782–204790.
11 Wang Y., Chang Y., Alkhateeb A. F., Alotaibi N. D. (2021), Adaptive fuzzy output-feedback tracking control for switched nonstrict-feedback nonlinear systems with prescribed performance, Circuits, Systems and Signal Processing, 40, 88–113.
12 Jiang B., Reza Karimi H. (2020), Further criterion for stochastic stability analysis of semi-Markovian jump linear systems, International Journal of Robust and Nonlinear Control, 30, 2689–2700.
13 Wang B., Zhu Q. (2020), Stability analysis of discrete-time semi-Markov jump linear systems, IEEE Transactions on Automatic Control, 65 (12), 5415–5421.
14 Hu W., Zhu Q., Reza Karimi H. (2019), Some improved Razumikhin stability criteria for impulsive stochastic delay differential systems, IEEE Transactions on Automatic Control, 64, 5207–5213.
15 Wang H., Zhu Q. (2020), Global stabilization of a class of stochastic nonlinear time-delay systems with siss inverse dynamics, IEEE Transactions on Automatic Control, 65 (10), 4448–4455.
16 Yuan C., Song S., Gao Q., Reza Karimi H., Pekar L., Guo S. (2020), A novel frequency-domain approach for the exact range of imaginary spectra and the stability analysis of LTI systems with two delays, IEEE Access, 8, 36595–36601.
17 Zhu Q. (2018), Stability analysis of stochastic delay differential equations with Levy noise, Systems & Control Letters, 118, 62–68.
18 Zhu Q., Huang T. (2020), Stability analysis for a class of stochastic delay nonlinear systems driven by G-Brownian motion, Systems & Control Letters, 140, 104699.
19 Li M., Chen H., Li X. (2020), Synchronization analysis of complex dynamical networks subject to delayed impulsive disturbances, Complexity, 2020, 1–12.
20 Zhu Q. (2019), Stabilization of stochastic nonlinear delay systems with exogenous disturbances and the event-triggered feedback control, IEEE Transactions on Automatic Control, 64, 3764–3771.
21 Azizi A. (2020), A case study on designing a sliding mode controller to stabilize the stochastic effect of noise on mechanical structures: residential buildings equipped with ATMD, Complexity, 2020, 1–17.
22 Ji W., Qiu J., Reza Karimi H. (2020), Fuzzy-model-based output feedback sliding mode control for discrete-time uncertain nonlinear systems, IEEE Transactions on Fuzzy Systems, 28 (8), 1519–1530.

23 Xiao B., Yang X., Reza Karimi H., Qiu J. (2019), Asymptotic tracking control for a more representative class of uncertain nonlinear systems with mismatched uncertainties, IEEE Transactions on Industrial Electronics, 66, 9417–9427.

24 Zhu Q., Wang H. (2018), Output feedback stabilization of stochastic feedforward systems with unknown control coefficients and unknown output function, Automatica, 87, 166–175.

25 Martinez D. I., Rubio J. J., Aguilar A., Pacheco J., Gutierrez G. J., Garcia V., Vargas T. M., Ochoa G., Cruz D. R., Juarez C. F. (2020), Stabilization of two electricity generators, Complexity, 2020, 1–13.

8

A Regulator to Attenuate the Perturbations and to Regulate the Rotatory Inverted Pendulum

8.1 Introduction

The perturbation attenuation problem is an important topic of the regulator processes, both in theory and in applications. This problem has been studied in many processes, more perturbations have been found and, at the same time, more strategies are proposed to attenuate these perturbations. Some examples of the aforementioned facts are [1–4].

The Chen et al. [1] investigate the problem of state-feedback robust stabilization for discrete-time fuzzy singularly perturbed processes. In [2], the attitude regulator of a quadrotor aircraft subject to a class of perturbations is recommended. In [3], a regulator to compensate the perturbations of a robotic manipulator is designed. In [4], a strategy which employs state-feedback integral regulator to achieve nonzero perturbation attenuation and zero steady-state error is proposed.

There exist some works for stabilization of pendulums, as are [5–10].

In [5], a Furuta pendulum is proposed. In [6], the stability of an inverted pendulum is recommended. The Hamiltonian regulator for stabilization of a rigid body process of two torques is developed in [7]. In [8], the Lagrangian regulator is recommended for stabilization of a cart inverted pendulum. In [9], a regulator based on the Lyapunov function is discussed for stabilization of a Furuta pendulum. In [10], a nonlinear regulator is designed for stabilization of a Furuta pendulum. In [11], a total energy regulator is applied to the cart inverted pendulum and Furuta pendulum.

Sometimes, not only the attenuate of the perturbations is required, if the process is unstable, a regulator to regulate the process would be required too. There are some works of pole placement regulator, but they are only applied to warrant the stability of the process. It is not common to see a regulator that regulates and attenuates the perturbations at the same time.

Perturbations in Mechatronic, Robotic, and Energy Systems: Estimation and Attenuation,
First Edition. Jose de Jesus Rubio. Published 2026 by John Wiley & Sons, Inc.

This chapter studies the structure at infinite regulator [12–14], the pole placement regulator [15], and the combination of both strategies to reach the stability and the attenuation of the perturbations.

This chapter is organized as follows: Section 8.2, presents the rotatory inverted pendulum process. Section 8.3 presents the problem of regulation using the pole placement strategy. Section 8.4 presents the problem of the perturbation attenuation. Section 8.5 presents the structure at infinite regulator to attenuate the perturbations. Section 8.6 applies both strategies to regulate and to attenuate the perturbations in a rotatory inverted pendulum process. Section 8.6 describes the simulations. Section 8.7 presents the conclusion.

8.2 Rotary Inverted Pendulum

Figure 8.1 shows the rotatory inverted pendulum.

The equations which describe the rotatory inverted pendulum are [16]:

$$\begin{aligned} C_1\ddot{\theta} + C_2\ddot{\alpha} &= T - B_\theta\dot{\theta}, \\ C_3\ddot{\alpha} + C_2\ddot{\theta} - C_4\alpha &= -B_\phi\dot{\alpha}, \end{aligned} \tag{8.1}$$

where $\alpha = \frac{\pi}{2} - \phi$ is the position of the pendulum, $\dot{\alpha}$ is the velocity of the pendulum, θ is the position of the arm, $\dot{\theta}$ is the velocity of the arm, $C_1 = m_1 l_{1c}^2 + I_{1z1} + m_2 l_1^2 + I_{2y2}$, $C_2 = m_2 l_1 l_{2c}$, $C_3 = m_2 l_{2c}^2 + I_{2y2}$, $C_4 = m_2 g l_{2c}$ are constants, m_2 is the pendulum mass, l_{2c} is the length of the center of mass for the pendulum, l_2 is the length of the pendulum, I_{2y2} is the inertia moment over y axis in the pendulum, g is the gravity, m_1 is the arm mass, l_{1c} is the length of the center of mass for the

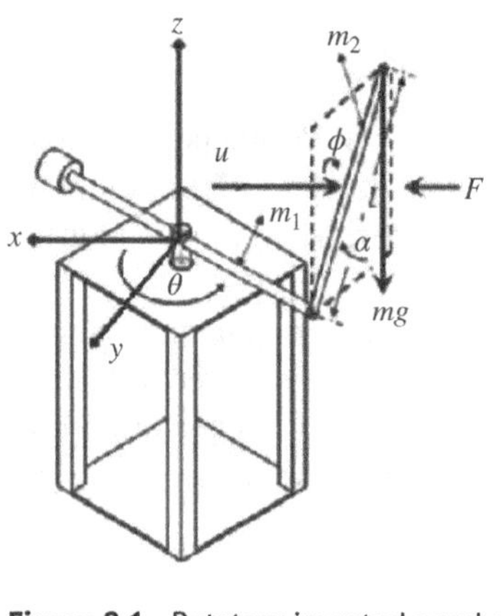

Figure 8.1 Rotatory inverted pendulum.

arm, l_2 is the length of the arm, I_{1z1} is the inertia moment over z axis in the arm, T is the torque of the motor, B_θ is the viscous damping constant of θ joint, and B_ϕ is the viscous damping constant of the ϕ joint.

Multiplying the first equation by C_3 and the second equation by C_2 of (8.1) gives the following equation:

$$\begin{aligned} &C_1C_3\ddot{\theta} + C_3C_2\ddot{\alpha} = C_3T - C_3B_\theta\dot{\theta} \\ &\Rightarrow C_3C_2\ddot{\alpha} + C_2^2\ddot{\theta} - C_2C_4\alpha = -C_2B_\phi\dot{\alpha}. \end{aligned} \tag{8.2}$$

Subtracting the second equation to the first gives:

$$\ddot{\theta}[C_1C_3 - C_2^2] = [-C_2C_4]\alpha + C_3T - C_3B_\theta\dot{\theta} + C_2B_\phi\dot{\alpha}. \tag{8.3}$$

Considering $-C_3B_\theta\dot{\theta} + C_2B_\phi\dot{\alpha} = 0$, gives the equation for the position of the arm of the rotatory inverted pendulum as follows:

$$\ddot{\theta}[C_1C_3 - C_2^2] = [-C_2C_4]\alpha + C_3T. \tag{8.4}$$

Multiplying the first equation by C_2 and the second equation by C_1 of (8.1) gives the following equation:

$$\begin{aligned} &C_1C_3\ddot{\alpha} + C_1C_2\ddot{\theta} - C_1C_4\alpha = C_1B_\phi\dot{\alpha} \\ &\Rightarrow C_1C_2\ddot{\theta} + C_2^2\ddot{\alpha} = -C_2T - C_2B_\theta\dot{\theta}. \end{aligned} \tag{8.5}$$

Subtracting the second equation to the first gives:

$$\ddot{\alpha}[C_1C_3 - C_2^2] = [C_1C_4]\alpha - C_2T + C_1B_\phi\dot{\alpha} + C_2B_\theta\dot{\theta}.$$

Considering $C_1B_\phi\dot{\alpha} + C_2B_\theta\dot{\theta} = 0$, gives the equation of the position of the pendulum of the rotatory inverted pendulum as follows:

$$\ddot{\alpha}[C_1C_3 - C_2^2] = [C_1C_4]\alpha - C_2T. \tag{8.6}$$

The modified equations of the rotatory inverted pendulum equations are (8.4) and (8.6).

Considering the states as $\chi_1 = \alpha$, $\chi_2 = \dot{\alpha}$, $\chi_3 = \theta$, $\chi_4 = \dot{\theta}$, the input as $\nu = T \in \Re^{1\times 1}$, and the output as $y = \alpha + \theta$ gives the following state space process:

$$\begin{aligned} &\dot{\chi} = A\chi + B\nu, \\ &y = C\chi, \end{aligned} \tag{8.7}$$

where $\chi = [\chi_1, \chi_2, \chi_3, \chi_4]^T \in \Re^{4\times1}$, $C = \begin{bmatrix} c_a & 0 & c_b & 0 \end{bmatrix} \in \Re^{1\times4}$,

$$A = \begin{bmatrix} 0 & 1 & 0 & 0 \\ \frac{c_1c_4}{c_1c_3 - c_2^2} & 0 & 0 & 0 \\ 0 & 0 & 0 & 1 \\ -\frac{c_2c_4}{c_1c_3 - c_2^2} & 0 & 0 & 0 \end{bmatrix} \in \Re^{4\times4}, \text{ and } B = \begin{bmatrix} 0 & -\frac{c_2}{c_1c_3 - c_2^2} & 0 & \frac{c_3}{c_1c_3 - c_2^2} \end{bmatrix}^T \in R^{4\times1}.$$

Table 8.1 shows the values of the constant parameters of the pendulum.

From Table 8.1 and considering $g = 9.81\,\text{m/s}^2$, gives $\frac{c_1c_4}{c_1c_3 - c_2^2} = 145.2$, $-\frac{c_2c_4}{c_1c_3 - c_2^2} = -29.19$, $-\frac{c_2}{c_1c_3 - c_2^2} = -487.57$, $\frac{c_3}{c_1c_3 - c_2^2} = 182.57$, $c_a = 1$, and $c_b = 1$, so the state space process of the rotatory inverted pendulum is rewritten as follows:

$$\begin{aligned} \dot{\chi} &= A\chi + B\nu, \\ y &= C\chi, \end{aligned} \tag{8.8}$$

where $C = \begin{bmatrix} 1 & 0 & 1 & 0 \end{bmatrix}$, $A = \begin{bmatrix} 0 & 1 & 0 & 0 \\ 145.2 & 0 & 0 & 0 \\ 0 & 0 & 0 & 1 \\ -29.19 & 0 & 0 & 0 \end{bmatrix}$, and

$B = \begin{bmatrix} 0 & -487.57 & 0 & 182.57 \end{bmatrix}^T$.

Table 8.1 Parameters of the rotatory inverted pendulum.

Parameter	Value	Parameter	Value
m_2	0.040691 kg	m_1	0.136729 kg
l_{2c}	0.15 m	l_{1c}	0.2 m
l_2	0.3 m	l_1	0.4 m
I_{2y2}	0.201×10^{-9}	I_{1z1}	-0.127×10^{-8}

8.3 The Problem of Regulation for the Pole Placement Strategy

First, the stability of the process is analyzed as follows [15]:

$$\begin{aligned} &\det(CI - A) = 0 \\ &\Rightarrow P(s) = s^n + a_{n-1}s^{n-1} + \cdots + a_1 s + a_0 = 0, \end{aligned} \tag{8.9}$$

where Ps is the characteristic polynomial of the process, a_0, and $a_1,\ldots, a_n$ are constant parameters. From (8.9), the eigenvalues are obtained. If the real part of all the eigenvalues are negative, the poles are located on the left-hand side of the imaginary plane and the process is stable, on the other case, the process is unstable.

If the process is unstable, a regulator can be used to force the real part of the eigenvalues to be negative.

The regulator is defined as [15]:

$$\nu = F\chi, \tag{8.10}$$

where $F \in \Re^{1\times 4}$ is the proportional gain of the regulator. Substituting (8.10) into (8.8) gives:

$$\dot{\chi} = A_\nu \chi, \tag{8.11}$$

where $A_\nu = A + BF$. Now, the stability of the regulation process is reached as follows:

$$\begin{aligned} &\det(CI - A_\nu) = 0, \\ &P_\nu = s^n + a_{n-1}s^{n-1} + \cdots + a_1 s + a_0 = 0, \\ &P_d = s^n + d_{n-1}s^{n-1} + \cdots + d_1 s + d_0 = 0, \\ &\Rightarrow P_\nu = P_d, \end{aligned} \tag{8.12}$$

where $P\nu$ is the characteristic polynomial of the regulation process, where Pd is the desired polynomial, a_0, $a_1,\ldots, a_n$, d_0, and $d_1,\ldots, d_n$ are constant parameters. From (8.12), the eigenvalues of $P\nu$ and Pd are obtained. Pd is selected such as the real part of all the eigenvalues is negative, and as $P\nu$ is forced to be equal to Pd, the real part of the eigenvalues $P\nu$ are forced to be negative too.

8.4 The Problem of Perturbation Attenuation

Let us consider the following linear process (A, B, C) given as:

$$\begin{aligned} &\dot{\chi}(t) = A\chi(t) + B\nu(t), \\ &y(t) = C\chi(t), \end{aligned} \tag{8.13}$$

where $t \geq 0$, $\chi(t) \in \mathcal{X} \approx \mathbb{R}^n$ is the state, $\nu(t) \in \mathcal{U} \approx \mathbb{R}^m$ is the regulator function, and $y(t) \in \mathcal{Y} \approx \mathbb{R}^p$ is the output. A, B, and C are the matrices, which represent the linear applications $\mathcal{A} : \mathcal{X} \to \mathcal{X}, \mathcal{B} : \mathcal{U} \to \mathcal{X}, \mathcal{C} : \mathcal{X} \to \mathcal{Y}$, respectively.

The perturbations process (A, B, C, E) is:

$$\begin{aligned} \dot{\chi}(t) &= A\chi(t) + B\nu(t) + Ed(t), \\ y(t) &= C\chi(t), \end{aligned} \tag{8.14}$$

where $t \geq 0$, $d(t) \in \mathcal{D} \approx \mathbb{R}^p$ are the perturbations and E is the matrix, which represents the linear application $\mathcal{E} : \mathcal{D} \to \mathcal{X}$.

Let us consider the measures of the states $\chi(t)$ and the perturbations $d(t)$ are known.

Theorem 8.1: *The perturbation attenuation problem has a solution if there exist a state feedback $\mathcal{F} : \mathcal{X} \to \mathcal{U}$ and a function $\mathcal{G} : \mathcal{D} \to \mathcal{U}$ such that the transfer function $d \to y$ is null because of the following regulator function:*

$$\nu = \mathcal{F}\chi + \mathcal{G}d, \tag{8.15}$$

where the transfer function of the output and the perturbations is:

$$G(s) = C(sI - A - BF)^{-1}(BG + E). \tag{8.16}$$

Proof. See [17] for the proof. ■

8.5 Stable Regulator Which Attenuate the Perturbations for the Rotatory Inverted Pendulum

Let us define the perturbations process:

$$\begin{aligned} \dot{\chi} &= A\chi + B\nu + Ed, \\ y &= C\chi, \end{aligned} \tag{8.17}$$

where A, B, C, χ, ν, and y are defined in (8.7) and (8.8);
$E = \begin{bmatrix} 0 & -3182.72 & 0 & 1186.71 \end{bmatrix}^T$; $d \in \Re$ are the perturbations.

8.5.1 Stability Analysis of the Process

The stability analysis is obtained using (8.9) as follows [15]:

$$\det\left(\begin{bmatrix} s & 0 & 0 & 0 \\ 0 & s & 0 & 0 \\ 0 & 0 & s & 0 \\ 0 & 0 & 0 & s \end{bmatrix} - \begin{bmatrix} 0 & 1 & 0 & 0 \\ 145.2 & 0 & 0 & 0 \\ 0 & 0 & 0 & 1 \\ -29.19 & 0 & 0 & 0 \end{bmatrix}\right) = 0 \tag{8.18}$$
$$\Rightarrow Ps = s^4 - 145.2s^2 = s^2(s - 145.2)(s + 145.2) = 0.$$

The roots of the polynomial Ps are $s = 0, s = 0, s = 145.2$, and $s = -145.2$. The roots of the polynomial are known as the eigenvalues of the process. As one of the eigenvalues is positive, the process is unstable, and a regulator should be proposed to force the process to become stable.

8.5.2 Controllability of the Process

The controllability of the process is analyzed as follows [15]:

$$C = \begin{bmatrix} B & AB & A^2B & A^3B \end{bmatrix}$$
$$\Rightarrow C = \begin{bmatrix} 0 & -487.57 & 0 & -70795. \\ -487.57 & 0 & -70795. & 0 \\ 0 & 182.57 & 0 & 14232. \\ 182.57 & 0 & 14232. & 0 \end{bmatrix}. \tag{8.19}$$

As the rank of matrix C is equal to 4, the process is controllable, so a regulator can be proposed.

8.5.3 Regulator Design

The regulator function of (8.15) is used, but F will be obtained with (8.10) to regulate the process and G will be obtained with (8.16) to attenuate the perturbations.

Considering $F = \begin{bmatrix} f_1 & f_2 & f_3 & f_4 \end{bmatrix}$ of (8.10) and using (8.12), Pv and Pd are given as:

$$\begin{aligned} Pv &= s^4 + (489.57f_2 - 182.57f_4)s^3 \\ &+ (489.57f_1 - 145.2 - 182.57f_3)s^2 + 12219f_4 s + 12219f_3, \\ Pd &= (s + 10)(s + 20)(s + 30)(s + 40) \\ &= s^4 + 100s^3 + 3500s^2 + 50000s + 240000. \end{aligned} \tag{8.20}$$

Considering $P\nu = Pd$ of (8.12) gives:

$$\begin{aligned} &F = \begin{bmatrix} f_1 & f_2 & f_3 & f_4 \end{bmatrix}, \\ &f_1 = 14.771, f_2 = 2.0268, \\ &f_3 = 19.642, f_4 = 4.0920. \end{aligned} \tag{8.21}$$

Let F of (8.21) and if $G = -6.5$, then the transfer function $G(s)$ of the output y and the perturbations d in (8.16) is:

$$G(s) = \frac{-13.51s^2 + 5090.7s + 12401}{s^4 + 241.13s^3 + 3470.6s^2 + 50275s + 2.4133 \times 10^5}. \tag{8.22}$$

Obtaining the convergence at infinite of $G(s)$ gives:

$$\begin{aligned} &\lim_{s\to\infty} G(s) \\ &= \lim_{s\to\infty} \frac{-13.51s^2 + 5090.7s + 12401}{s^4 + 241.13s^3 + 3470.6s^2 + 50275s + 2.4133 \times 10^5}. \end{aligned} \tag{8.23}$$

Multiplying $G(s)$ by 1 expressed as $\left(\frac{s^{-4}}{s^{-4}}\right)$ gives:

$$\begin{aligned} &\lim_{s\to\infty} G(s) \\ &= \lim_{s\to\infty} \frac{-13.51s^2 + 5090.7s + 12401}{s^4 + 241.13s^3 + 3470.6s^2 + 50275s + 2.4133 \times 10^5} \left(\frac{s^{-4}}{s^{-4}}\right) \\ &= \lim_{s\to\infty} \frac{-13.51s^{-2} + 5090.7s^{-3} + 12401s^{-4}}{1 + 241.13s^{-1} + 3470.6s^{-2} + 50275s^{-3} + 2.413\,3 \times 10^5 s^{-4}} \\ &= \frac{0}{1} = 0. \end{aligned} \tag{8.24}$$

As $G(s)$ using F and G is not zero, the perturbations are not attenuated, but as $\lim_{s\to\infty} G(s)$ using F and G is zero, the perturbations are attenuated.

Finally, the obtained regulator function (8.15) for the rotatory inverted pendulum model (8.8) is:

$$\nu = \begin{pmatrix} 14.771 & 2.0268 & 19.642 & 4.0920 \end{pmatrix} \chi - 6.5d. \tag{8.25}$$

8.6 Simulation

In this section, the simulations are recommended. The perturbations are selected as $d = \sin(8\pi t)$. Using (8.18), the rotatory inverted pendulum is unstable. The stable regulator of (8.10) and (8.21), and the stable regulator that attenuate the perturbations of (8.15) and (8.25), will be compared for the rotatory inverted pendulum process.

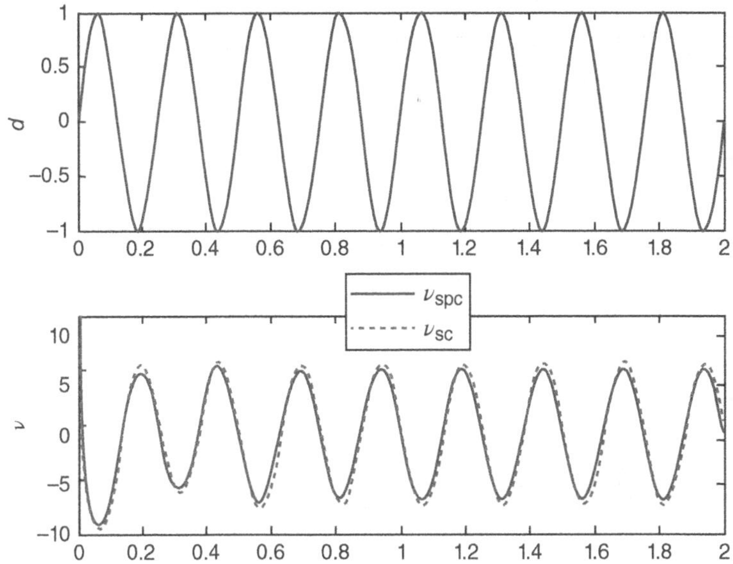

Figure 8.2 Perturbations and regulator functions.

Figure 8.2 shows the perturbations named d, the regulator function of the stable regulator named ν_{sc}, and the stable regulator function, which attenuate the perturbations named ν_{spc}.

Figure 8.3 shows the states of the rotatory inverted pendulum process of (8.8) using the stable regulator named χ_{sc} and the stable regulator, which attenuate the perturbations named χ_{spc}.

Figure 8.4 shows the outputs of the rotatory inverted pendulum process using the stable regulator named y_{sc} and the stable regulator, which attenuate the perturbations named y_{spc}.

From Figures 8.2 to 8.4, it can be seen that the states and the output of the rotatory inverted pendulum with the stable regulator have perturbations, but the states and the output of the rotatory inverted pendulum with stable regulator that attenuate the perturbations do not have perturbations.

8.7 Concluding Remarks

The chapter recommended two strategies for the regulation and the attenuation of perturbations for a rotatory inverted pendulum process. The strategies were the pole placement regulator and the structure at infinite regulator. In the future, the perturbations will be considered unknown and will be attenuated or attenuated using the evolving processes.

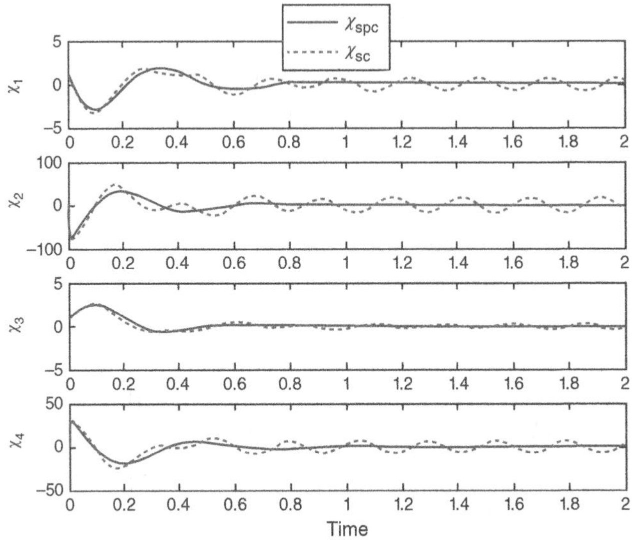

Figure 8.3 States of the rotatory inverted pendulum.

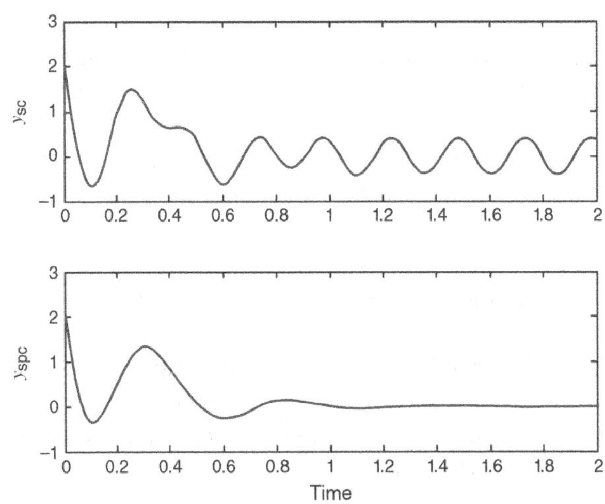

Figure 8.4 Output of the rotatory inverted pendulum.

References

1 Chen J., Sun F., Yin Y., Hu C. (2011), State feedback robust stabilisation for discrete-time fuzzy singularly perturbed systems with parameter uncertainty, IET Control Theory and Applications, 5 (10), 1195–1202.

2 Zhang R., Quan Q., Cai K.-Y. (2011), Attitude control of a quadrotor aircraft subject to a class of time-varying disturbances, IET Control Theory and Applications, 5 (9), 1140–1146.

3 Dunnigan M. W., Wronka C. M. (2011), Comparison of control techniques for a robotic manipulator with base disturbances, IET Control Theory and Applications, 5 (8), 999–1012.

4 Li H., Sun Z., Liu H., Sun F., Deng J. (2011), State feedback integral control of networked control systems with external disturbance, IET Control Theory and Applications, 5 (2), 283–290.

5 Furuta K., Okutani T., Sone H. (1978), Computer control of a double inverted pendulum, Computers & Electrical Engineering, 5 (1), 67–84.

6 Quintero P., Silva J., Alejandro H. (2009), Estabilidad de un péndulo con forzamiento periódico arbitrario, Revista Mexicana de Fisica, 55 (2), 161–167.

7 Aguilar C., Suarez M. S., Guzman F. (2008), A Hamiltonian control approach for the stabilization of the angular velocity of a rigid body system controlled by two torques, Revista Mexicana de Fisica, 54 (4), 265–270.

8 Aguilar C., Gutierrez O., Sossa H. (2008), Controlled Lagrangian approach to the stabilization of the inverted pendulum system, Revista Mexicana de Fisica, 54 (4), 329–335.

9 Aguilar C., Sossa J. H. (2007), Stabilization of the Furuta pendulum based on a Lyapunov function, Nonlinear Dynamics, 49, 1–8.

10 Aguilar C., Suarez M. S., Gutierres O. O. (2010), The direct Lyapunov method for the stabilisation of the Furuta pendulum, International Journal of Control, 83 (11), 2285–2293.

11 Viola G., Ortega R., Banavar R., Acosta J. A., Astolfi A. (2007), Total energy shaping control of mechanical systems: simplifying the matching equations via coordinate changes, IEEE Transactions on Automatic Control, 52 (6), 1093–1099.

12 Rabah R., Malabre M. (1999), On the structure at infinity of linear delay systems with application to the disturbance decoupling problem, Kybernetika, 35 (6), 668–680.

13 Rabah R., Malabre M. (1993), Structure at infinity, model matching and disturbance rejection for linear systems with delays, Kybernetika, 29 (5), 485–498.

14 Malabre M., Kucera V. (1984), Infinite structure and exact model matching problem: a geometric approach. IEEE Transactions on Automatic Control, 3, 266–268.

15 Chen C.-T. (1999), Linear Systems, Theory and Design, third edition, ISBN 0-19-511-777-8.

16 Awtar S., King N., Allen T., Bang I., Hagan M., Skidmore D., Craig K., (2002), Inverted pendulum systems: rotatory and arm-driven-a mechatronic system design case study, Mechatronics, 12, 357–370.

17 Rubio J. J., Figueroa M., Perez Cruz J. H., Rumbo J. (2012), Control to stabilize and mitigate disturbances in a rotary inverted pendulum, Revista Mexicana de Fisica E, 58 (2), 107–112.

9

Proportional-derivative Regulator with Inverse Dead Zone for Pendulum Processes

9.1 Introduction

Non-smooth nonlinear benefits such as dead zone, backlash, and hysteresis are common in actuators and sensors such as mechanical connections, hydraulic servo-valves, and electric servomotors; they also appear in biomedical processes. Dead zone is one of the most important non-smooth nonlinearities in many industrial processes, which can severely limit the system performance; and its study has been drawing much interest in the regulation community for a long time [1].

There is some research about regulation processes. In [2], the regulation of the inverted-car pendulum is recommended. The regulation of the Furuta pendulum is recommended in [3]. In [4], the dissipative regulation problem is investigated for a class of discrete time-varying processes. The distributed H∞ filtering problem for a class of nonlinear processes is considered in [5]. The recursive finite-horizon filtering problem for a class of nonlinear time-varying processes is addressed in [6, 7]. In [8], the authors present a solution to the problem of the quadratic min–max regulator for polynomial uncertain processes. The problem of a two-player differential game affected by matched uncertainties with only the output measurement available for each player is considered by [9]. The stability analysis and H∞ regulator for a class of discrete-time switched linear parameter-varying processes is concerned in [10]. The sliding mode regulation problem for uncertain nonlinear discrete-time stochastic processes with H_2 performance constraints is designed in [11]. In [12], the sliding mode regulation problem is considered for discrete-time processes. In [13], the description about the modeling and regulation of wind turbine system is addressed. A model to describe the dynamics of a homogeneous viscous fluid in an open pipe is recommended in [14]. In [15], the authors consider the problems of robust stability and H∞ regulation for a class of networked regulation processes with longtime delays. The H∞ regulator issue for a class of networked regulation processes with packet dropouts and time-varying delays is recommended in [16]. From the above studies, in [2, 3, 8, 9, 13, 14], the authors propose proportional-derivative (PD) regulators; however, none considers processes with dead-zone inputs.

There is some work about the regulation of processes with dead-zone inputs. In [17–22], the authors recommended the regulation of nonlinear processes with dead-zone inputs. Nevertheless, they do not research about the pendulum processes. The pendulum mathematical models have different structures with respect to the nonlinear processes addressed in the above studies; so, a new design may be developed.

In this chapter, a PD regulator with inverse dead zone is recommended for the regulation of pendulum processes with dead-zone inputs. One main benefit of this chapter is that the pendulum mathematical model is rewritten as a robotic mathematical model to satisfy a property. Later, the property is applied to warrant the stability of the recommended regulator.

The chapter is organized as follows: In Section 9.2, the mathematical model of the robotic arm with dead-zone inputs is recommended. In Section 9.3, the mathematical model of the pendulum processes with dead-zone inputs is recommended. In Section 9.4, PD regulator with inverse dead zone is recommended. In Section 9.5, the recommended strategy is used for the regulation of two pendulum processes. Section 9.6 presents conclusions and suggests future research directions.

9.2 Mathematical Model of the Robotic Arms with Dead Zone Inputs

The main concern of this section is to understand some concepts of robot dynamics. The equation of motion for the constrained robotic manipulator with n degrees of freedom, considering the contact force and the constraints, is given in the joint space as follows:

$$M(q)\ddot{q} + C(q, \dot{q})\dot{q} + G(q) = \tau, \tag{9.1}$$

where $q \in \Re^{n\times1}$ is the joint angles or link displacements of the manipulator, $M(q) \in \Re^{n\times n}$ is the robot inertia matrix, which is symmetric and positive definite, $C(q, \dot{q}) \in \Re^{n\times n}$ contains the centripetal and Coriolis terms and $G(q)$ denotes the gravity terms, and τ is the dead-zone output. The nonsymmetric dead zone can be represented by:

$$\tau = \mathrm{DZ}(u) = \begin{cases} m_r\,(u - b_r) & u \geq b_r \\ 0 & b_l < u < b_r, \\ m_l\,(u - b_l) & u \leq b_l \end{cases} \tag{9.2}$$

where m_r and m_l are the right and left constant slopes for the dead-zone benefit, b_r and b_l represent the right and left breakpoints. Note that u is the input of the dead zone and the regulator input of the global system.

Define the following two states as follows:

$$\begin{aligned}\chi_1 &= q \in \Re^{n\times 1},\\ \chi_2 &= \dot{q} \in \Re^{n\times 1},\\ \nu &= \tau \in \Re^{n\times 1},\end{aligned} \tag{9.3}$$

where $\chi_1 = \begin{bmatrix}\chi_{11} & \chi_{12}\end{bmatrix}^T = \begin{bmatrix}q_1 & q_2\end{bmatrix}^T$, $\chi_2 = \begin{bmatrix}\chi_{21} & \chi_{22}\end{bmatrix}^T = \begin{bmatrix}\dot{q}_1 & \dot{q}_2\end{bmatrix}^T$ for $n = 2$. So (9.1) can be rewritten as:

$$\begin{aligned}&\dot{\chi}_1 = \chi_2,\\ &M(\chi_1)\dot{\chi}_2 + C(\chi_1, \chi_2)\chi_2 + G(\chi_1) = \nu,\end{aligned} \tag{9.4}$$

where $M(\chi_1)$, $C(\chi_1, \chi_2)$, and $G(\chi_1)$ are described in (9.1), the dead zone ν is [17, 19, 20, 22]:

$$\nu = \mathrm{DZ}(u) = \begin{cases} m_r\,(u - b_r) & u \geq b_r \\ 0 & b_l < u < b_r. \\ m_l\,(u - b_l) & u \leq b_l \end{cases} \tag{9.5}$$

The parameters m_r, m_l, b_r, and b_l are described in (9.2), and u is the regulator input of the system. Figure 9.1 shows the dead zone [17].

Property 1. The inertia matrix is symmetric and positive definite, i.e., [23–25]:

$$m_1|\chi|^2 \leq \chi^T M(\chi_1)\chi \leq m_2|\chi|^2,$$

where m_1 and m_2 are known positive scalar constants, $\chi = [\chi_1, \chi_2]^T$.

Property 2. The centripetal and Coriolis matrix is skew-symmetric, i.e., satisfies the following relationship [23–25]:

$$\chi^T\left[\dot{M}(\chi_1) - 2C(\chi_1, \chi_2)\right]\chi = 0, \tag{9.6}$$

where $\chi = [\chi_1, \chi_2]^T$.

The normal PD regulator is:

$$\nu = -K_p\widetilde{\chi}_1 - K_d\widetilde{\chi}_2, \tag{9.7}$$

where $\widetilde{\chi}_1 = \chi_1 - \chi_1^d$ and $\widetilde{\chi}_2 = \chi_2 - \chi_2^d$, K_p and K_d are positive definite, symmetric, and constant matrices.

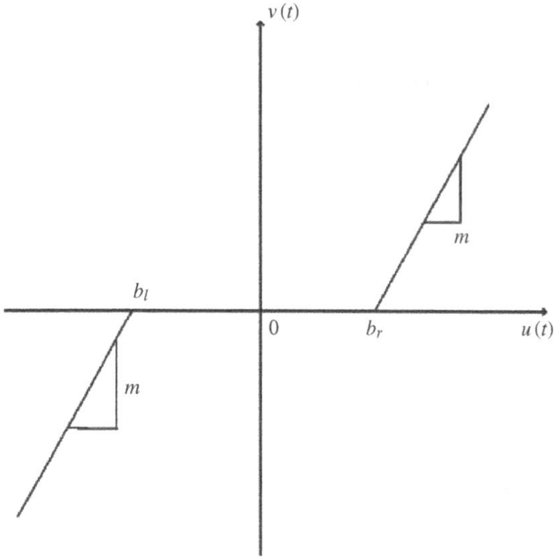

Figure 9.1 The dead zone.

9.3 Mathematical Model of the Pendulum Processes with Dead Zone Inputs

The mathematical model of the pendulum processes can be rewritten as the mathematical model of the robotic arms; however, Property 2 is not directly satisfied. Pendulum mathematical models are rewritten as the robotic mathematical models because in this chapter, if the above sentence is true, Property 2 of the robotic processes can be used to warrant the stability of the regulator applied to the pendulum processes. The following lemmas let us to adapt Property 2 to the pendulum processes.

Lemma 9.1: *A pendulum model can be rewritten as a robotic arm model (9.4). Nevertheless, it cannot satisfy Property 2.*

Proof. See [26] for the proof. ∎

Lemma 9.2: *Pendulum model (9.4) can be rewritten as follows:*

$$\begin{aligned} \dot{\chi}_1 &= \chi_2, \\ M(\chi_1)\dot{\chi}_2 &= -\widehat{C}(\chi_1, \chi_2)\chi_2 - N(\chi_1, \chi_2) + \nu, \end{aligned} \tag{9.8}$$

where

$$M(\chi_1) = \begin{bmatrix} m_{11} & m_{12} \\ m_{21} & m_{22} \end{bmatrix},$$
$$\widehat{C}(\chi_1, \chi_2) = \begin{bmatrix} \widehat{c}_{11} & \widehat{c}_{12} \\ \widehat{c}_{21} & \widehat{c}_{22} \end{bmatrix}, \tag{9.9}$$
$$N(\chi_1, \chi_2) = \begin{bmatrix} n_1 \\ n_2 \end{bmatrix}.$$

The input ν is given by (9.5), where $c_{11} = \widehat{c}_{11} + \overline{c}_{11}$, $c_{12} = \widehat{c}_{12} + \overline{c}_{12}$, $c_{21} = \widehat{c}_{21} + \overline{c}_{21}$, $c_{22} = \widehat{c}_{22} + \overline{c}_{22}$, $n_1 = g_1 + \overline{c}_{11}\chi_{21} + \overline{c}_{12}\chi_{22}$ and $n_2 = g_2 + \overline{c}_{21}\chi_{21} + \overline{c}_{22}\chi_{22}$, where χ_1 *and χ_2 are defined in (9.3), and the following modified property is satisfied:*

$$\chi^T \left[\dot{M}(\chi_1) - 2\widehat{C}(\chi_1, \chi_2) \right] \chi = 0. \tag{9.10}$$

Proof. See [26] for the proof. ■

In the following section, a stable regulator for the pendulum processes will be designed.

9.4 Proportional Derivative Regulator with Inverse Dead Zone

The regulation case is considered in this chapter, i.e., the desired velocity is $\chi_2^d = 0$. The PD regulator with inverse dead zone u is as follows:

$$u = \mathrm{DZ}^{-1}(\nu_{pd}) = \begin{cases} \frac{1}{m_r}\nu_{pd} + b_r & \nu_{pd} > 0 \\ 0 & \nu_{pd} = 0, \\ \frac{1}{m_l}\nu_{pd} + b_l & \nu_{pd} < 0 \end{cases} \tag{9.11}$$

where the parameters m_r, m_l, b_r, and b_l are defined as in (9.2); the auxiliary PD regulator is:

$$\nu_{pd} = -K_p\widetilde{\chi}_1 - K_d\widetilde{\chi}_2 + \widehat{N} - K\mathrm{sign}(\widetilde{\chi}_2), \tag{9.12}$$

where $\widetilde{\chi}_1 = \chi_1 - \chi_1^d$ is the tracking error, χ_1 and $\widetilde{\chi}_2 = \chi_2$ are defined in (9.3). Here, $\chi_1^d = \left[\chi_{11}^d \quad \chi_{12}^d\right] \in \mathfrak{R}^{n\times1}$ is the desired position, K_p and K_d are positive definite, and $\widehat{N} \in \mathfrak{R}^{n\times1}$ is an approximation of $N(\chi_1, \chi_2)$, where $N(\chi_1, \chi_2) \in \mathfrak{R}^{n\times1}$ are

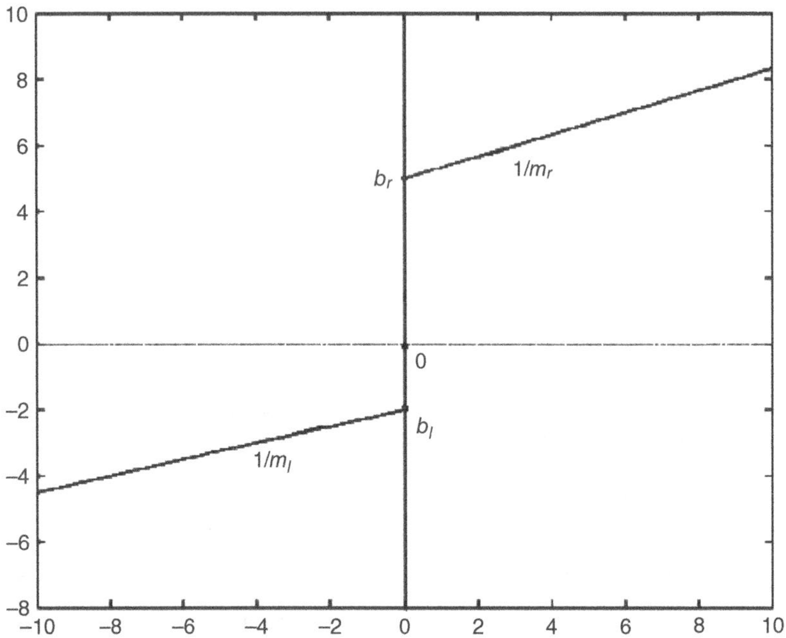

Figure 9.2 The inverse dead zone.

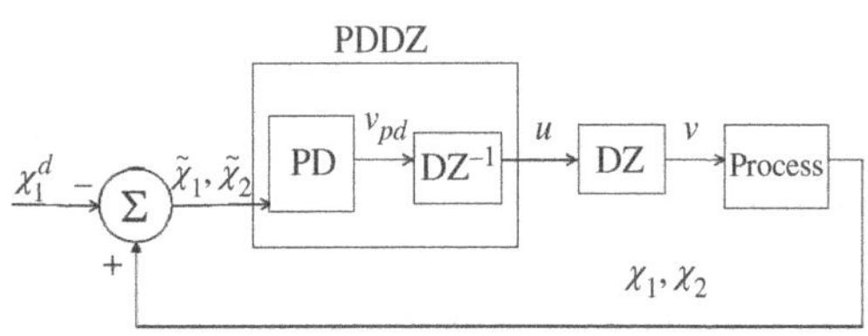

Figure 9.3 The recommended regulator.

the nonlinear terms of (9.8). Figure 9.2 shows the inverse dead zone [17, 22] and Figure 9.3 shows the recommended regulator named PD with inverse dead zone (PDDZ). It is considered that the approximation error $\widetilde{N} = \widehat{N} - N(\chi_1, \chi_2)$ is bounded as:

$$\left|\widetilde{N}\right| \leq \overline{N}. \tag{9.13}$$

Now the convergence of the closed-loop system is discussed in Theorem 9.1.

Theorem 9.1: *The error of the close-loop system with the PD regulator (9.11) and (9.12) for the pendulum processes with dead-zone inputs (9.5) and (9.8), is asymptotically stable, and the error of the velocity parameter $\widetilde{\chi}_2$ will converge to:*

$$\lim_{T\to\infty} \sup \|\widetilde{\chi}_2\|^2 = 0, \tag{9.14}$$

where T is the final time, $\widetilde{\chi}_2 = \chi_2$, $\overline{N} \leq K$, $K_p > 0$, and $K_d > 0$.

Proof. See [26] for the proof. ■

Remark 9.1: *The recommended regulator is used for the regulation case, i.e., the desired velocity is $\chi_2^d = 0$. The general case when $\chi_2^d \neq 0$ is not considered in this chapter.*

9.5 Simulations

In this section, the PDDZ will be compared with the PD regulator with gravity compensation of [23], named PD, for the regulation of two pendulum processes with dead-zone inputs. In this chapter, the root mean square error (RMSE) [1, 27] is used for the comparison results, and it is given as:

$$\text{RMSE} = \left(\frac{1}{T} \int_0^T \widetilde{\chi}^2 dt \right)^{\frac{1}{2}}, \tag{9.15}$$

where $\widetilde{\chi}^2 = \widetilde{\chi}_{12}^2$ or $\widetilde{\chi}^2 = v_1^2$.

9.5.1 Example 1

Consider the inverted-car pendulum [2] of Figure 9.4.

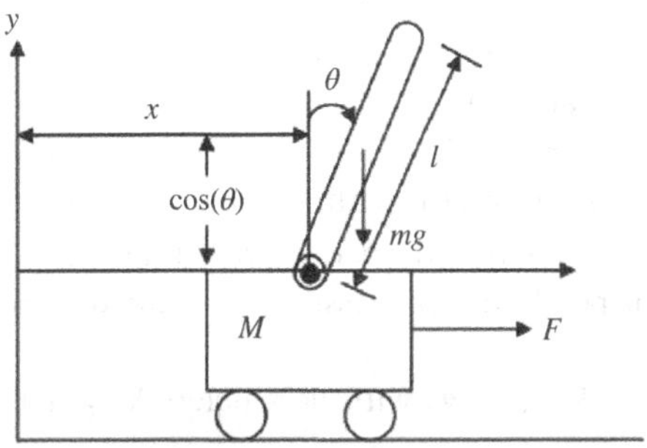

Figure 9.4 Inverted-car pendulum.

Inverted-car pendulum is written as (9.1), and it is detailed as follows:

$$M(\chi_1) = \begin{bmatrix} m_{11} & m_{12} \\ m_{21} & m_{22} \end{bmatrix},$$
$$C(\chi_1, \chi_2) = \begin{bmatrix} c_{11} & c_{12} \\ c_{21} & c_{22} \end{bmatrix}, \tag{9.16}$$
$$G(\chi_1) = \begin{bmatrix} g_1 \\ g_2 \end{bmatrix},$$

where $m_{11} = M + m$, $m_{12} = m_{21} = ml\cos(\chi_{12})$, $m_{33} = I + ml^2$, $c_{12} = -ml \sin(\chi_{12})\chi_{22}$, the other parameters of $C(\chi_1, \chi_2)$ are zero, $g_2 = -mgl\sin(\chi_{12})$, and the other parameter of $G(\chi_1)$ is zero. $\sin(\cdot)$ is the sine function, $\cos(\cdot)$ is the cosine function, $I = 0.5$ kg m^2 is the pendulum inertia, $M = 0.136729$ kg is the mass of the car, $m = 0.040691$ kg is the pendulum mass, $l = 0.15$ m is the pendulum length, $\chi_{12} = \theta$ is the angle respect to the y axis, $\nu = F$ is the motion force of the car, $\chi_{11} = x$ is the motion distance of the car, and $g = 9.81$ m/s^2 is the constant acceleration due to gravity. It can be proven that Property 2 of (9.6) is not satisfied.

Inverted-car pendulum is written as (9.8), and it is detailed as follows:

$$M(\chi_1) = \begin{bmatrix} m_{11} & m_{12} \\ m_{21} & m_{22} \end{bmatrix},$$
$$\widehat{C}(\chi_1, \chi_2) = \begin{bmatrix} \widehat{c}_{11} & \widehat{c}_{12} \\ \widehat{c}_{21} & \widehat{c}_{22} \end{bmatrix}, \tag{9.17}$$
$$N(\chi_1, \chi_2) = \begin{bmatrix} n_1 \\ n_2 \end{bmatrix},$$

where $m_{11} = M + m$, $m_{12} = m_{21} = ml\cos(\chi_{12})$, $m_{33} = I + ml^2$, $\widehat{c}_{12} = \widehat{c}_{21} = -\frac{1}{2}ml\sin(\chi_{12})\chi_{22}$, the other parameters of $\widehat{C}(\chi_1, \chi_2)$ are zero, $n_1 = -\frac{1}{2}ml\sin(\chi_{12})\chi_{22}$, $n_2 = -mgl\sin(\chi_{12}) + \frac{1}{2}ml\sin(\chi_{12})\chi_{22}$; so, it can be proven that the property of the lemma of (9.10) is satisfied.

PDDZ is given by (9.11) and (9.12) as $\nu_{pd} = -K_p\widetilde{\chi}_{12} - K_d\widetilde{\chi}_{22} + \widehat{n}_1 - K\text{sign}(\widetilde{\chi}_{22})$ with parameters $K_p = 200$, $K_d = 20$, $\widehat{N} = \widehat{n}_1$, $\widehat{n}_1 = 0.01$, with $K = 0.01$. Conditions given in (9.14) $\overline{N} \leq K$, $K_p > 0$, and $K_d > 0$ are satisfied; so, the error of the closed-loop dynamics of the PDDZ applied for pendulum processes is warranted to be asymptotically stable.

PD is given by [23] as $\nu_{pd} = -K_p\widetilde{\chi}_{12} - K_d\widetilde{\chi}_{22} + g_2$ with parameters $K_p = 400$, $K_d = 20$ and $G(\chi_1) = g_1$.

Comparison results for the regulator functions are shown in Figure 9.5, position states are shown in Figure 9.6, and comparison results for the regulator errors are shown in Figure 9.7. Comparison of the square norm of the velocity errors

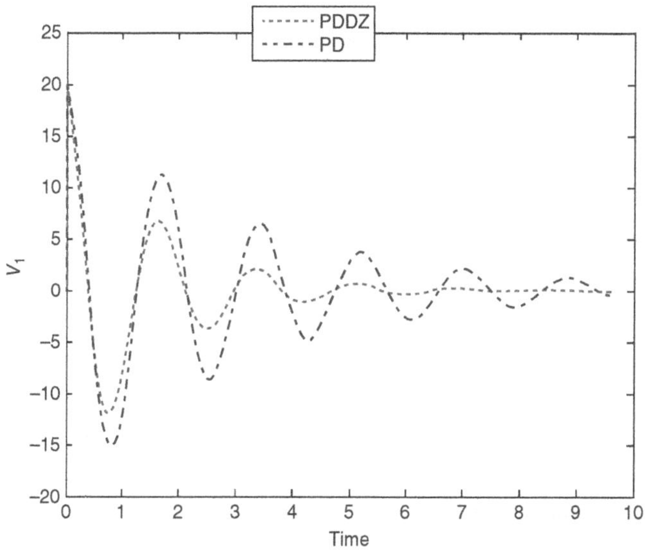

Figure 9.5 Input for Example 1.

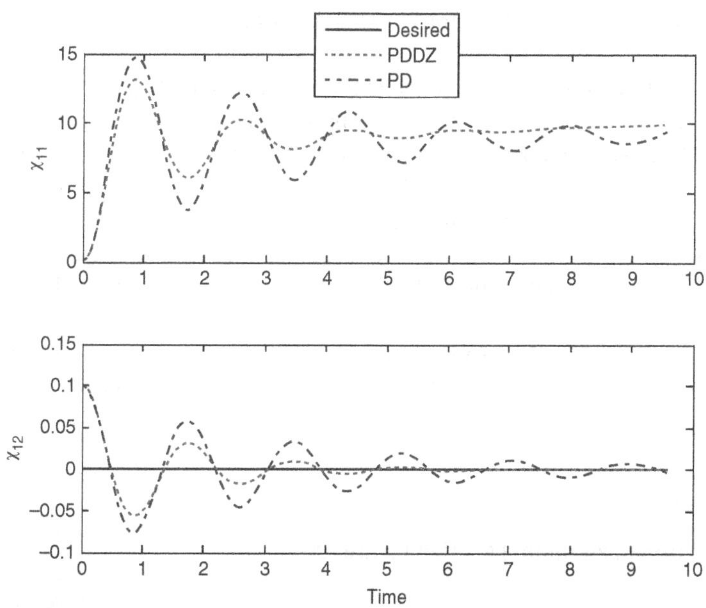

Figure 9.6 Position states for Example 1.

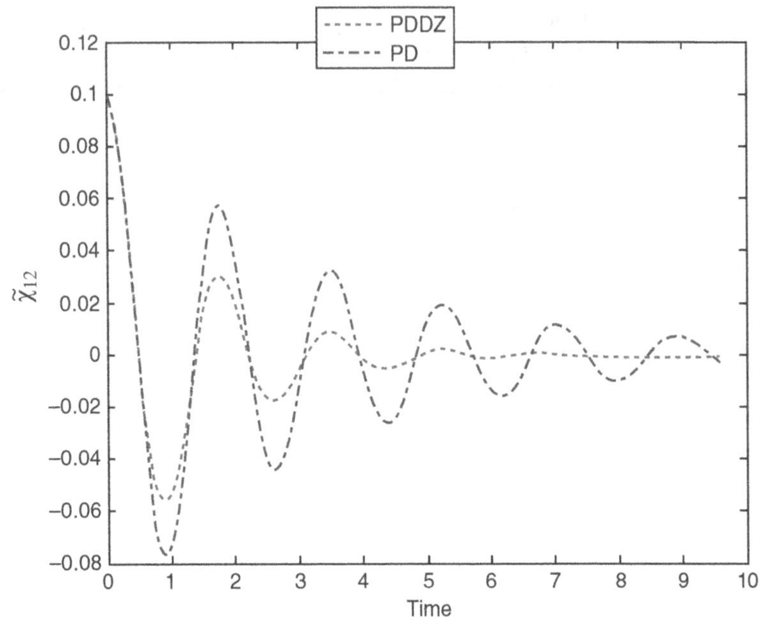

Figure 9.7 Position state error for Example 1.

$\|\tilde{\chi}_2\|^2$ of (9.14) for the regulators is recommended in Figure 9.8. From theorem of (9.14), $\|\tilde{\chi}_2\|^2$ will converge to zero for the PDDZ. Table 9.1 shows the RMSE results using (9.15).

The most important variable to regulate is the pendulum angle $\chi_{12} = \theta$. This variable may reach zero even if it starts with another value, as in this example. Note that the PD strategy requires bigger gains than the PDDZ strategy to obtain satisfactory results. From Figures 9.5 to 9.7, it can be seen that the PDDZ improves the PD because the signal of the plant for the first follows the desired signal better than the second, and in the first, the inputs are smaller than in the second. From Figure 9.8, it is shown that the PDDZ improves the PD because the velocity error $\|\tilde{\chi}_2\|^2$ recommended by the first is smaller than that recommended by the second. From Table 9.1, it can be shown that the PDDZ achieves better accuracy when compared with the PD because the RMSE is smaller for the first than for the second.

9.5.2 Example 2

Consider the Furuta pendulum [3, 27] of Figure 9.9.

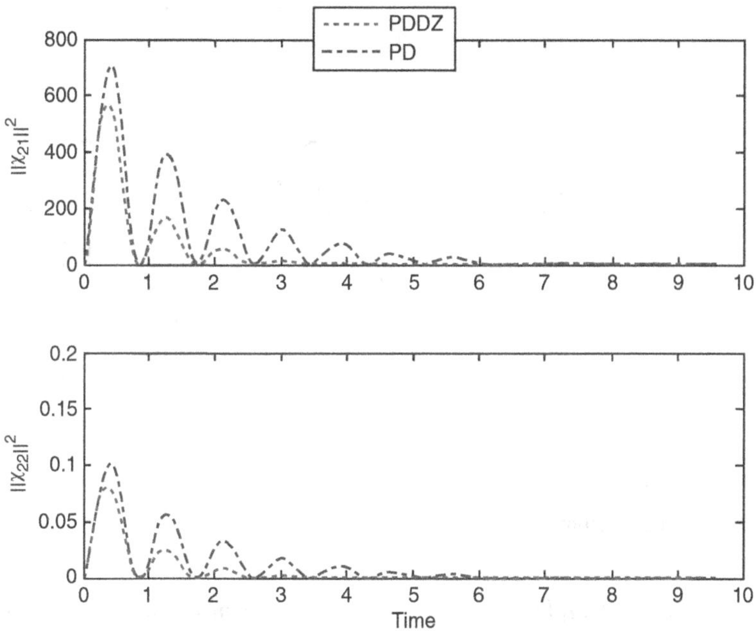

Figure 9.8 Velocity errors for Example 1.

Table 9.1 Error results for Example 1.

Strategies	**RMSE for $\widetilde{\chi}$**	**RMSE for ν**
PD	4.2487×10^{-4}	15.5044
PDDZ	2.1995×10^{-4}	7.7534

Furuta pendulum is written as (9.1), and it is detailed as follows:

$$
\begin{aligned}
M(\chi_1) &= \begin{bmatrix} m_{11} & m_{12} \\ m_{21} & m_{22} \end{bmatrix}, \\
C(\chi_1, \chi_2) &= \begin{bmatrix} c_{11} & c_{12} \\ c_{21} & c_{22} \end{bmatrix}, \\
G(\chi_1) &= \begin{bmatrix} g_1 \\ g_2 \end{bmatrix},
\end{aligned}
\tag{9.18}
$$

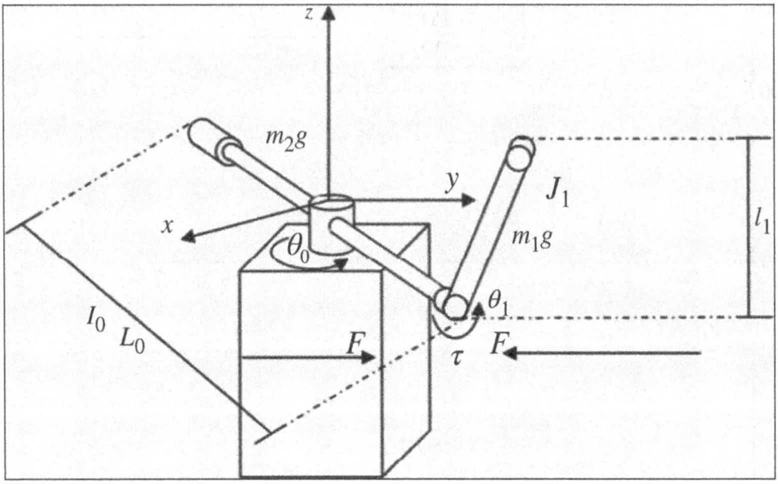

Figure 9.9 Furuta pendulum.

where $m_{11} = I_0 + m_1\left(L_0^2 + l_1^2\sin^2(\chi_{12})\right)$, $m_{12} = m_{21} = m_1 l_1 L_0 \cos(\chi_{12})$, $m_{33} = J_1 + m_1 l_1^2$, $c_{11} = m_1 l_1^2 \sin(2\chi_{12})\chi_{22}$, $c_{12} = -m_1 l_1 L_0 \sin(\chi_{12})\chi_{22}$, $c_{21} = -m_1 l_1^2 \sin(\chi_{12})\cos(\chi_{12})\chi_{21}$, $g_2 = -m_1 g l_1 \sin(\chi_{12})$, and the other parameter of $G(\chi_1)$ is zero. $\sin(\cdot)$ is the sine function, $\cos(\cdot)$ is the cosine function, $I_0 = 0.5$ kg m^2 is the arm inertia, $J_1 = 0.5$ kg m^2 is the pendulum inertia, $m_2 = 0.34$ kg is the arm mass, $m_1 = 0.24$ kg is the pendulum mass, $L_0 = 0.293$ m is the arm length, $l_1 = 0.28$ m is the pendulum length, $\chi_{11} = \theta_0$ is the arm angle, $\chi_{12} = \theta_1$ is the pendulum angle, $\nu = F$ is the motion torque of the arm, and g = 9.81 m/s^2 is the constant acceleration due to gravity. It can be proven that Property 2 of (9.6) is not satisfied.

Furuta pendulum is written as (9.8), and it is detailed as follows:

$$
\begin{aligned}
M(\chi_1) &= \begin{bmatrix} m_{11} & m_{12} \\ m_{21} & m_{22} \end{bmatrix}, \\
\widehat{C}(\chi_1, \chi_2) &= \begin{bmatrix} \widehat{c}_{11} & \widehat{c}_{12} \\ \widehat{c}_{21} & \widehat{c}_{22} \end{bmatrix}, \\
N(\chi_1, \chi_2) &= \begin{bmatrix} n_1 \\ n_2 \end{bmatrix},
\end{aligned}
\tag{9.19}
$$

where $m_{11} = M + m$, $m_{12} = m_{21} = ml\cos(\chi_{12})$, $m_{33} = I + ml^2$, $\widehat{c}_{11} = m_1 l_1^2 \sin(\chi_{12})\cos(\chi_{12})\chi_{22}$, $\widehat{c}_{12} = -\frac{1}{2} m_1 l_1 L_0 \sin(\chi_{12})\chi_{22}$, $\widehat{c}_{21} = -\frac{1}{2} m_1 l_1^2 \sin(\chi_{12})\cos(\chi_{12})\chi_{22}$, and the other parameter of $\widehat{C}(\chi_1, \chi_2)$ is zero. $n_1 = m_1 l_1^2 \sin(\chi_{12})\cos(\chi_{12})\chi_{22} - \frac{1}{2} m_1 l_1 L_0 \sin(\chi_{12})\chi_{22}$ and $n_2 = -m_1 g l_1 \sin(\chi_{12}) - \frac{1}{2} m_1 l_1^2 \sin(\chi_{12})\cos(\chi_{12})(\chi_{21} - \chi_{22})$; so, it can be proven that the property of the lemma of (9.10) is satisfied.

PDDZ is given by (9.11) and (9.12) as $\nu_{pd} = -K_p\widetilde{\chi}_{12} - K_d\widetilde{\chi}_{22} + \widehat{n}_1 - K\text{sign}(\widetilde{\chi}_{22})$ with parameters $K_p = 200$, $K_d = 20$, $\widehat{N} = \widehat{n}_1$, $\widehat{n}_1 = 0.01$, with $K = 0.01$. Conditions given in (9.14) $\overline{N} \leq K$, $K_p > 0$, and $K_d > 0$ are satisfied; so, the error of the closed-loop dynamics of the PDDZ applied for pendulum processes is warranted to be asymptotically stable.

PD is given by [23] as $\nu_{pd} = -K_p\widetilde{\chi}_{12} - K_d\widetilde{\chi}_{22} + g_2$ with parameters $K_p = 400$, $K_d = 20$, and $G(\chi_1) = g_2$.

Comparison results for the regulator functions are shown in Figure 9.10, position states are shown in Figure 9.11 and comparison results for the regulator errors are shown in Figure 9.12. Comparison of the square norm of the velocity errors $||\widetilde{\chi}_2||^2$ of (9.14) for the regulators is recommended in Figure 9.13. From theorem of (9.14), $||\widetilde{\chi}_2||^2$ will converge to zero for the PDDZ. Table 9.2 shows the RMSE results using (9.15).

The most important variable to regulate is the pendulum angle $\chi_{12} = \theta_1$. This variable may reach zero even if it starts with another value, as in this example.

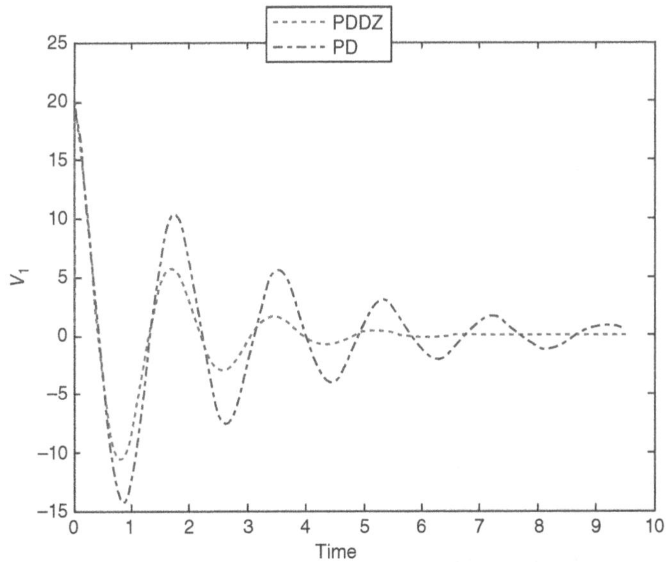

Figure 9.10 Input for Example 2.

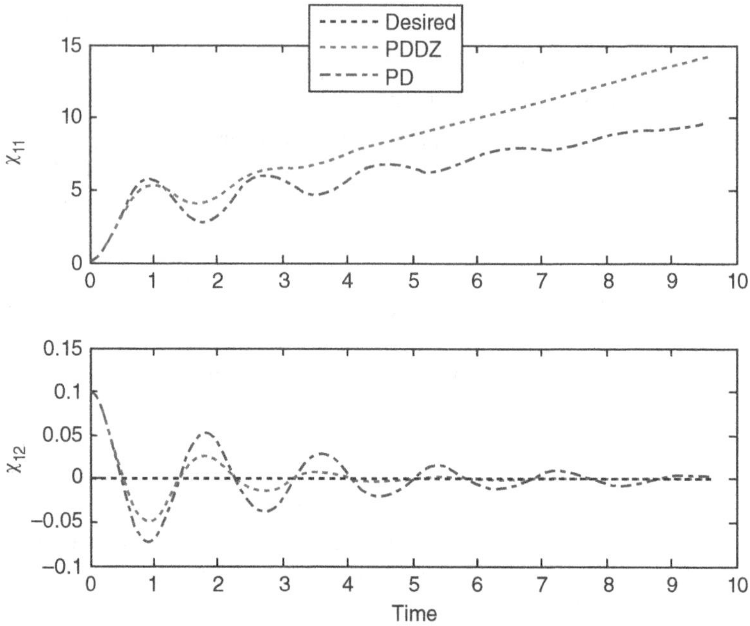

Figure 9.11 Position states for Example 2.

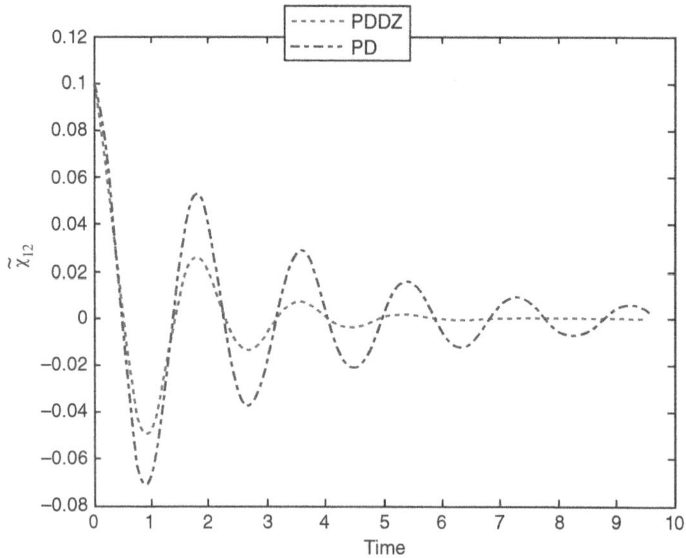

Figure 9.12 Position state error for Example 2.

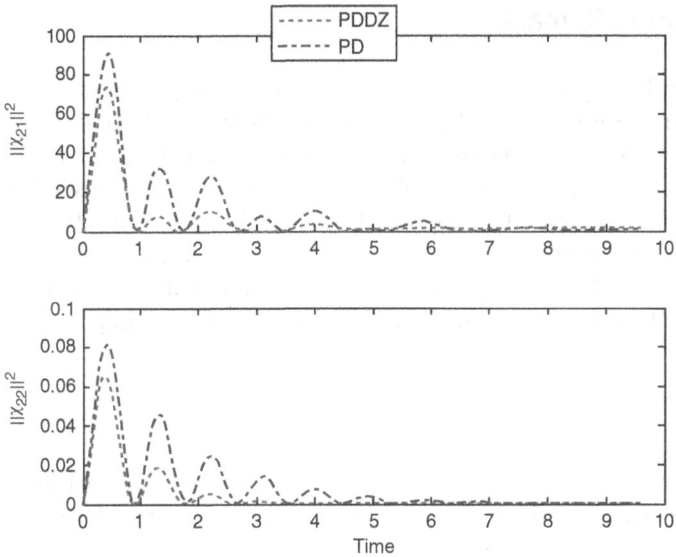

Figure 9.13 Velocity errors for Example 2.

Table 9.2 Error results for Example 2.

Strategies	RMSE for $\tilde{\chi}$	RMSE for ν
PD	3.7760×10^{-4}	13.6043
PDDZ	2.0158×10^{-4}	6.8328

Note that the PD strategy requires bigger gains than the PDDZ strategy to obtain satisfactory results. From Figures 9.10 to 9.12, it can be seen that the PDDZ improves the PD because the signal of the plant for the first follows the desired signal better than the second, and in the first, the inputs are smaller than in the second. From Figure 9.13, it is shown that the PDDZ improves the PD because the velocity error $||\tilde{\chi}_2||^2$ recommended by the first is smaller than that recommended by the second. From Table 9.2, it can be shown that the PDDZ achieves better accuracy when compared with the PD because the RMSE is smaller for the first than for the second.

9.6 Concluding Remarks

In this chapter, a PD regulator with inverse dead zone for pendulum processes with dead-zone inputs is recommended. The simulations showed that the recommended strategy achieves better performance when compared with the PD regulator with gravity compensation for the regulation of two pendulum processes. The results illustrate the viability, efficiency, and the potential of the approach, especially important in pendulum processes. As a future research, the recommended study will be improved considering that some parameters of the regulator are unknown, or it will consider the communication delays and packet dropout.

References

1 Rubio J. J., Perez-Cruz J. H. (2014), Evolving intelligent system for the modelling of nonlinear systems with dead-zone input, Applied Soft Computing, 14 (Part B), 289–304.
2 Aguilar-Ibáñez C. F., Martinez García J. C., Lopez-Soria A., Rubio J. J. (2011), On the stabilization of the inverted-cart pendulum using the saturation function, Mathematical Problems in Engineering, 2011, 1–14.
3 Aguilar-Ibañez C., Suarez-Castañon M. S., Gutierres-Frias O. O. (2010), The direct Lyapunov method for the stabilisation of the Furuta pendulum, International Journal of Control, 83, (111), 2285–2293.
4 Ding D., Wang Z., Hu J., Shu H. (2013), Dissipative control for state-saturated discrete time-varying systems with randomly occurring nonlinearities and missing measurements, International Journal of Control, 86 (4), 674–688.
5 Dong H., Wang Z., Lam J., Gao H. (2014), Distributed filtering in sensor networks with randomly occurring saturations and successive packet dropouts, International Journal of Robust and Nonlinear Control, 24 (12), 1743–1759.
6 Hu J., Wang Z., Shen B., Gao H. (2013), Quantised recursive filtering for a class of nonlinear systems with multiplicative noises and missing measurements, International Journal of Control, 86 (4), 650–663.
7 Hu J., Wang Z., Shen B., Gao H. (2013), Gain-constrained recursive filtering with stochastic nonlinearties and probabilistic sensor delays, IEEE Transactions on Signal Processing, 61 (5), 1230–1238.
8 Jimenez-Lizarraga M., Basin M., Rodriguez-Ramirez P. (2012), Robust mini-max regulator for uncertain non-linear polynomial systems, IET Control Theory & Applications, 6 (7), 963–970.
9 Ferreira de Loza A., Jimenez-Lizarraga M., Fridman L. (2012), Robust output Nash strategies based on sliding mode observation in a two-player differential game, Journal of the Franklin Institute, 349 (4), 1416–1429.

10 Lu Q., Zhang L., Reza Karimi H., Shi Y. (2013), H∞ control for asynchronously switched linear parameter-varying systems with mode-dependent average dwell time, IET Control Theory & Applications, 7 (5), 677–683.

11 Ma L. F., Wang Z. D., Guo Z. (2009), Robust H_2 sliding mode control for non-linear discrete-time stochastic systems, IET Control Theory & Applications, 3 (11), 1537–1546.

12 Niu Y., Ho D. W. C., Wang Z. (2010), Improved sliding mode control for discrete-time systems via reaching law, IET Control Theory & Applications, 4 (11), 2245–2251.

13 Soriano L. A., Yu W., Rubio J. J. (2013), Modeling and control of wind turbine, Mathematical Problems in Engineering, 2013, 1–32.

14 Rubio J. J., Ordaz G., Jimenez-Lizarraga M., Cabrera R. I., (2013), General solution to the Navier-Stokes equation to describe the dynamics of a homogeneous viscous fluid in an open pipe, Revista Mexicana de Fisica, 59 (3), 217–223.

15 Wang Y., Reza Karimi H., Xiang Z. (2013), Delay-dependent H∞ control for networked control systems with large delays, Mathematical Problems in Engineering, 2013, 1–10.

16 Wang Y., Reza Karimi H., Xiang Z. (2013), H∞ control for networked control systems with time delays and packet dropouts, Mathematical Problems in Engineering, 2013, 1–10.

17 Abrir S., Xie W. F., Su C. Y. (2007), Adaptive tracking of nonlinear systems with non-symmetric dead-zone input, Automatica, 43 (3), 522–530.

18 Abrir S. (2012), Invariant-manifold approach to the stabilization of feedforward nonlinear systems having uncertain dead-zone inputs, International Symposium of Mechatronics and its Applications, 2012,1–5.

19 Perez Cruz J. H., Ruiz Velazquez E., Rubio J. J., de Alva Padilla C. A. (2012), Robust adaptive neurocontrol of SISO nonlinear systems preceded by unknow deadzone, Mathematical Problems in Engineering, 2012, 1–22.

20 Perez-Cruz J. H., Rubio J. J., Ruiz-Velazquez E., Solis-Perales G. (2012), Tracking control based on recurrent neural networks for nonlinear systems with multiple inputs and unknown deadzone, Abstract and Applied Analysis, 2012, 1–18.

21 Wang Z., Zhang Y., Fang H. (2008), Neural adaptive control for a class of nonlinear systems with unknown deadzone, Neural Computing and Applications, 17 (4), 339–345.

22 Zhou J., Shen X. Z. (2007), Robust adaptive control of nonlinear uncertain plants with unknown dead-zone, IET Control Theory & Applications, 1 (1), 25–32.

23 Lewis F. L., Dawson D. M., Abdallah C. T. (2004), Control of Robot Manipulators, Theory and Practice, ISBN: 0-8247-4072-6.

24 Slotine J. E., Li W. (1991), Applied Nonlinear Control, Macmillan Pub. Co., Englewood Cliffs, NJ 07632.

25 Spong M. W., Vidyasagar M. (1989), Robot Dynamics and Control, John Wiley & Sons.

26 Rubio J. J., Zamudio Z., Pacheco J., Mujica Vargas D., (2013), Proportional derivative control with inverse dead-zone for pendulum systems, Mathematical Problems in Engineering, 2013, 1–9.

27 Rubio J. J., Figueroa M., Perez Cruz J. H., Rumbo J. (2012), Control to stabilize and mitigate disturbances in a rotary inverted pendulum, Revista Mexicana de Fisica E, 58 (2), 107–112.

10

Sliding-mode Regulator of Robotic Arms with Dead Zone

10.1 Introduction

Non-smooth nonlinear characteristics, such as dead zone, backlash, and hysteresis, commonly occur in actuators and sensors, including mechanical connections, hydraulic servo-valves, and electric servo-motors; these also appear in biomedical processes. The dead zone is one of the most important non-smooth nonlinearities in many industrial processes, which can severely limit the system performance; and its study has been drawing much interest in the regulation community for a long time. Consequently, the regulation issue in which the behaviors in the dead zone are unknown has become great interest. One alternative to solve this problem is the sliding-mode (SM) strategy.

There is some interesting work about the regulation of processes with dead zone is investigated. In [1, 2], the regulation of nonlinear processes with input dead zone is investigated. In [3], the tracking regulation of processes with dead zone is considered. The regulators of robotic processes with dead zone are recommended in [4–6]. In [7], the adaptive regulator of a system in the presence of dead zone is described. Tracking positioning in the presence of the dead zone of a robot manipulator actuator is studied in [8]. In [9], a robust compensation scheme for a robot manipulator with dead zone is recommended. An adaptive neural network tracking regulator of a robotic manipulator with dead zone is detailed in [10]. In [11], an approximation-based adaptive tracking regulator is introduced. The tracking regulator of an uncertain robot is recommended in [12]. An adaptive impedance regulator of a robotic manipulator is developed in [13]. The previous investigations show that the dead zone is a current and recent strategy.

There is some novel research about the SM regulator. The observer-based SM regulators are investigated in [14–16]. In [17–20], new SM regulators for delayed processes are recommended. The SM regulator of a stochastic system is explained in [21]. In [22, 23], the SM strategies in electric vehicles are addressed. An observer-based adaptive neural network regulator is addressed in [24]. In [25–27], the SM regulators of robotic processes are considered. The aforementioned research shows that the SM regulator is an interesting and novel strategy.

The SM regulator has two problems: (1) the produced chattering, because it can damage the devices, and (2) the SM surface, because it requires many derivatives of the system. This chapter considers a design regulator where the knowledge of the behaviors in the dead zone and gravity are not required, i.e., the SM strategy is utilized in the regulator law instead of the unknown behaviors. The recommended strategy is focused on solving the mentioned problems as follows: (1) the chattering is decreased by the utilization of a smaller parameter in the SM regulator because of the combination with the proportional-derivative (PD) strategy, (2) the SM surface of the SM strategy is substituted by the velocity variable.

So, the introduced regulator is selected to achieve three main benefits: (1) it warrants its stability by the utilization of the Lyapunov strategy, (2) it provides an alternative for the compensation of the dead zone and gravity, and (3) it supposes an alternative for decreasing the chattering in the SM behavior.

The chapter is organized as follows. In Section 10.2, the mathematical model of robotic arms with dead zone and gravity is presented. In Section 10.3, the SM regulator for the regulation of robotic arms with unknown behaviors in the dead zone and gravity is introduced. In Section 10.4, the recommended strategy is applied for the regulation of two robotic arms. Section 10.5 presents conclusions and future research.

10.2 Mathematical Model of Robotic Arms with Dead Zone and Gravity

The objective of this section is to understand some concepts of robotic arms. The mathematical model for the constrained robotic arms with n degrees of freedom, considering the contact force and the constraints, is given in the joint space as follows:

$$M(p)\ddot{p} + C(p, \dot{p})\dot{p} + G(p) = f(p, \dot{p}, \ddot{p}) = \tau, \tag{10.1}$$

where $p \in \Re^{n\times1}$ is the position and $\dot{p} \in \Re^{n\times1}$ is the velocity of the joint angles or link displacements in the robotic arm, $M(p) \in \Re^{n\times n}$ is the robot inertia matrix, which is symmetric and positive definite, $C(p, \dot{p}) \in \Re^{n\times n}$ contains the centripetal and Coriolis terms, and $G(p)$ are the gravity terms, τ is the dead zone, and $f(p, \dot{p}, \ddot{p})$ is a nonlinear function, which describes the robot dynamics. The dead-zone terms are represented as follows [1, 2]:

$$\tau = \mathrm{DZ}(w) = \begin{cases} n_r\,(w - a_r) & w \geq a_r \\ 0 & a_l < w < a_r, \\ n_l\,(w - a_l) & w \leq a_l \end{cases} \tag{10.2}$$

where n_r and n_l are the right and left constant slopes for the dead-zone behavior, and a_r and a_l represent the right and left breakpoints. Note that w are the inputs of the dead zone and the inputs of the regulator system.

Define the two states and inputs as follows:

$$\begin{aligned} \chi_1 &= p \in \Re^{n\times 1}, \\ \chi_2 &= \dot{p} \in \Re^{n\times 1}, \\ \nu &= \tau \in \Re^{n\times 1}, \end{aligned} \tag{10.3}$$

where $\chi_1 = \begin{bmatrix} \chi_{11} & \chi_{12} \end{bmatrix}^T = \begin{bmatrix} p_1 & p_2 \end{bmatrix}^T$ and $\chi_2 = \begin{bmatrix} \chi_{21} & \chi_{22} \end{bmatrix}^T = \begin{bmatrix} \dot{p}_1 & \dot{p}_2 \end{bmatrix}^T$. So, the state-space system of (10.1) can be rewritten as follows:

$$\begin{aligned} \dot{\chi}_1 &= \chi_2, \\ M(\chi_1)\dot{\chi}_2 + C(\chi_1, \chi_2)\chi_2 + G(\chi_1) &= f(\chi_1, \chi_2, \dot{\chi}_2) = \nu, \end{aligned} \tag{10.4}$$

where $M(\chi_1)$, $C(\chi_1, \chi_2)$, $G(\chi_1)$, and $f(\chi_1, \chi_2, \dot{\chi}_2)$ are explained in (10.1), and the dead-zone terms ν of (10.2) are represented as follows:

$$\nu = \mathrm{DZ}(w) = \begin{cases} n_r\,(w - a_r) & w \geq a_r \\ 0 & a_l < w < a_r. \\ n_l\,(w - a_l) & w \leq a_l \end{cases} \tag{10.5}$$

The parameters n_r, n_l, a_r, and a_l are explained in (10.2), and w are the inputs of the regulator system. Figure 10.1 shows the dead-zone terms [1].

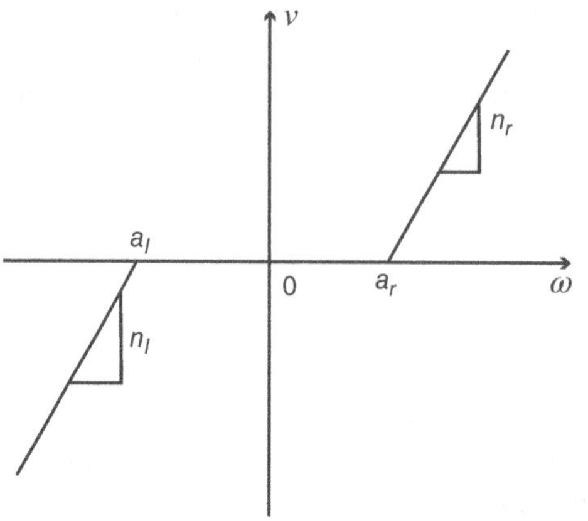

Figure 10.1 The dead zone.

The following two properties will be used in the next section to reach the stability of the recommended regulator.

Property 1. The inertia matrix is symmetric and positive definite [28, 29]:

$$m_m|\chi|^2 \leq \chi^T M(\chi_1)\chi \leq m_M|\chi|^2,$$

where m_m, m_M are known positive scalar constants, $\chi = [\chi_1, \chi_2]^T$, and $M(\chi_1)$ is defined in (10.4).

Property 2. The centripetal and Coriolis matrix is skew-symmetric, i.e., it satisfies the following relationship [28, 29]:

$$\chi^T \left[\dot{M}(\chi_1) - 2C(\chi_1, \chi_2)\right] \chi = 0, \tag{10.6}$$

where $\chi = [\chi_1, \chi_2]^T$, $M(\chi_1)$, and $C(\chi_1, \chi_2)$ are defined in (10.4).

The following two definitions describe the PD regulator with gravity compensation and the SM regulator because they will be utilized for the comparisons in Section 10.4.

Definition 1. The PD regulator with gravity compensation is as follows [28, 30]:

$$\nu = -K_p\widetilde{\chi}_1 - K_d\widetilde{\chi}_2 + G(\chi_1), \tag{10.7}$$

where $\widetilde{\chi}_1 = \chi_1 - \chi_1^d \in \Re^{n\times1}$, $\widetilde{\chi}_1 \in \Re^{n\times1}$ is the regulation error, χ_1 is the position, and $\chi_1^d \in \Re^{n\times1}$ is the constant desired position, $\widetilde{\chi}_2 = \chi_2$ is the velocity, and K_p and K_d are positive definite, symmetric, and constant matrices. Note that in the regulator of (10.7), the gravity terms $G(\chi_1)$ are totally known.

Definition 2. The SM regulator is as follows [28, 30]:

$$\nu = \frac{1}{n_r}\left[-K_p\widetilde{\chi}_1 - L_s\text{sign}\left(\eta_s\right)\right],$$
$$\eta_s = \widetilde{\chi}_2 + \lambda_s\widetilde{\chi}_1, \tag{10.8}$$

where $\widetilde{\chi}_1 = \chi_1 - \chi_1^d \in \Re^{n\times 1}$, $\widetilde{\chi}_1 \in \Re^{n\times 1}$ is the regulation error, χ_1 is the position and $\chi_1^d \in \Re^{n\times 1}$ is the constant desired position, $\widetilde{\chi}_2 = \chi_2$ is the velocity, and K_p and λ_s are positive definite, symmetric, and constant matrices. χ_1^d and χ_2^d are the desired references. L_s is a constant selected by the designer such as $\left|f(\chi_1, \chi_2, \dot{\chi}_2)\right| \leq L_s$, where $f(\chi_1, \chi_2, \dot{\chi}_2)$ is defined in (10.4). sign$(\cdot)$ is the signum function, η_s is the SM surface, and n_r is defined in (10.5).

10.3 Sliding-mode Regulator for the Regulation of Robotic Arms with Dead Zone and Gravity

From [8–10, 12, 13, 30], the actuators of robotic processes are symmetric, which is named $n_r = n_l$ in (10.5); so, this behavior can be rewritten as follows:

$$\begin{aligned}
\nu = \text{DZ}(w) &= \begin{cases} n_r\left(w - a_r\right) & w \geq a_r \\ 0 & a_l < w < a_r \\ n_r\left(w - a_l\right) & w \leq a_l \end{cases} \\
&= \begin{cases} n_r\left(w - a_r\right) & w \geq a_r \\ n_r\left(w - w\right) & a_l < w < a_r \\ n_r\left(w - a_l\right) & w \leq a_l \end{cases} \\
&= \begin{cases} n_r w & w \geq a_r \\ n_r w & a_l < w < a_r \\ n_r w & w \leq a_l \end{cases} + \begin{cases} -n_r a_r & w \geq a_r \\ -n_r w & a_l < w < a_r \\ -n_r a_l & w \leq a_l \end{cases} \\
&= n_r w - \begin{cases} n_r a_r & w \geq a_r \\ n_r w & a_l < w < a_r \\ n_r a_l & w \leq a_l \end{cases} \\
&= n_r w - d(w),
\end{aligned} \tag{10.9}$$

where $d(w) = \begin{cases} n_r a_r & w \geq a_r \\ n_r w & a_l < w < a_r \\ n_r a_l & w \leq a_l \end{cases}$ are considered as perturbations. It is important to note that the terms of $d(w)$ are bounded as follows:

$$|d(w)| \leq \overline{d}. \tag{10.10}$$

It is known that the gravity terms $G(\chi_1)$ of (10.11) are bounded as follows:

$$|G(\chi_1)| \leq \overline{G}. \tag{10.11}$$

The regulation case is considered in this study because the desired velocity states are assumed to be $\chi_2^d = 0$, and the desired references are constant. The SM regulator w is defined as follows:

$$\nu_{\text{ssm}} = w = \frac{1}{n_r}\left[-K_p\widetilde{\chi}_1 - K_d\widetilde{\chi}_2 - K\text{sign}(\widetilde{\chi}_2)\right], \tag{10.12}$$

where the parameter n_r is explained in (10.5), $\widetilde{\chi}_1 = \chi_1 - \chi_1^d \in \Re^{n\times 1}$, $\widetilde{\chi}_1 \in \Re^{n\times 1}$ is the regulation error, and $\chi_1^d \in \Re^{n\times 1}$ is the constant desired position. $\widetilde{\chi}_2 = \chi_2$, K_p, $K_d \in \Re^{n\times n}$ are positive definite, symmetric, and constant matrices. sign$(\cdot)$ is the signum function. K is a constant selected by the designer such as $\overline{d} + \overline{G} \leq K$, and $\overline{d}$ and $\overline{G}$ are defined in (10.10) and (10.11), respectively. It is important to note that the behaviors of $d(w)$ and $G(\chi_1)$ are unknown because only their upper bounds $\overline{d}$ and $\overline{G}$ are utilized.

Remark 10.1: The regulator of (10.10)–(10.12) is selected to find three main objectives: (1) to warrant its stability by the utilization of the Lyapunov strategy in the application of the processes with the form (10.4) and (10.5), (2) for the compensation of the unknown dead zone and gravity behaviors in the system (10.4) and (10.5), and (3) to suppose an alternative for the decreasing of the chattering in the SM behavior.

Figure 10.2 shows the SM regulator named stable sliding mode (SSM) in the application to robotic arms with unknown behavior in the dead zone and gravity named Process and DZ.

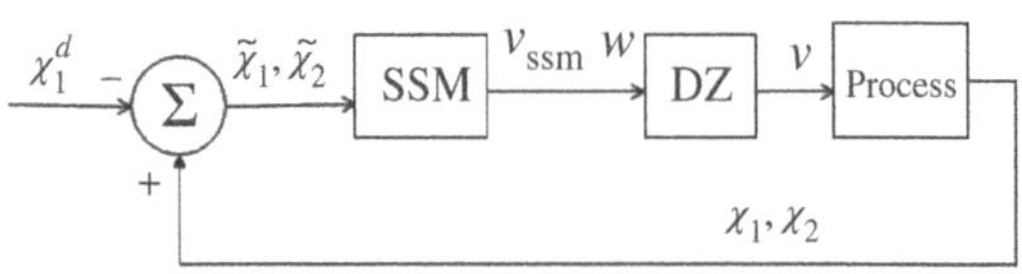

Figure 10.2 The SM regulator.

Remark 10.2: There are some works that consider the boundedness of some plant dynamics, such as the neural network regulator of [2], the observer-based SM regulators of [14–16], and the least square regulator of [30]. Taking into account these previous works, the Equations (10.10) and (10.11), which consider the boundedness of some plant dynamics, are correctly used. In addition, the dynamics are not required to be known because only their upper bounds are utilized.

Now, the convergence of the closed-loop system is discussed.

Theorem 10.1: *The regulation error in the close-loop system of the SM regulator (10.12) and robotic arms with unknown behaviors in the dead zone and gravity (10.4) and (10.5) is asymptotically stable, and the regulation error of the velocity states $\widetilde{\chi}_2$ will converge to:*

$$\lim_{T\to\infty} \sup \left\| \widetilde{\chi}_2 \right\|^2 = 0, \tag{10.13}$$

where T is the final time, $\widetilde{\chi}_2 = \chi_2$, $|d(w)| \leq \overline{d}$, $|G(\chi_1)| \leq \overline{G}$, and $\overline{d} + \overline{G} \leq K$.

Proof. See [31] for the proof. ■

Remark 10.3: There are some interesting works that consider the regulator design for the application to robotic processes such as [4–6, 8–10, 12, 13, 30]. The difference between the recommended strategy and the aforementioned research is that while they utilize intelligent strategies, this study employs the SM strategy.

10.4 Comparisons

In this section, the recommended SM regulator named SSM will be compared with the PD regulator with gravity compensation given in (10.7) named PD and with the SM regulator given in (10.8) named SM for the regulation of two robotic arms: the transelevator and articulated arms. The design objective of the regulators is that the trajectory of the states in robotic arms must follow the trajectories of desired constant references as fast as possible. The root mean square error (RMSE) [30] is employed for the comparisons, and it is given as follows:

$$\text{RMSE} = \left(\frac{1}{T} \int_0^T \widetilde{\chi}^2 dt \right)^{\frac{1}{2}}, \tag{10.14}$$

where $\widetilde{\chi}^2 = \widetilde{\chi}_{11}^2 + \widetilde{\chi}_{12}^2 + \cdots + \widetilde{\chi}_{1n}^2 + \widetilde{\chi}_{21}^2 + \widetilde{\chi}_{22}^2 + \cdots + \widetilde{\chi}_{2n}^2$ for the states or $\widetilde{\chi}^2 = v_1^2 + v_2^2 + \cdots + v_n^2$ for the inputs. $\widetilde{\chi}_{11}^2 = \left(\chi_{11} - \chi_{11}{}^d\right)^2$, $\widetilde{\chi}_{12}^2 = \left(\chi_{12} - \chi_{12}{}^d\right)^2$,

$\widetilde{\chi}_{1n}^2 = (\chi_{1n} - \chi_{1n}{}^d)^2$ are the regulation errors, $\widetilde{\chi}_{21}^2 = \chi_{21}^2$, $\widetilde{\chi}_{22}^2 = \chi_{22}^2$, $\widetilde{\chi}_{2n}^2 = \chi_{2n}^2$ are the velocity errors, $\nu_1^2 = \nu_1^2$, $\nu_2^2 = \nu_2^2$, $\nu_n^2 = \nu_n^2$ are the input errors, χ_{11}, χ_{12}, χ_{1n} are the positions, χ_{11}^d, χ_{12}^d, χ_{1n}^d are the constant desired positions, χ_{21}, χ_{22}, χ_{2n} are the velocities, and ν_1, ν_2, ν_3 are the inputs.

10.4.1 Transelevator

Consider the transelevator of Figure 10.3.

The transelevator is written as (10.1), and it is detailed as follows:

$$M(\chi_1) = \begin{bmatrix} m_{11} & m_{12} & m_{13} \\ m_{21} & m_{22} & m_{23} \\ m_{31} & m_{32} & m_{33} \end{bmatrix},$$
$$C(\chi_1, \chi_2) = \begin{bmatrix} c_{11} & c_{12} & c_{13} \\ c_{21} & c_{22} & c_{23} \\ c_{31} & c_{32} & c_{33} \end{bmatrix}, \tag{10.15}$$
$$G(\chi_1) = \begin{bmatrix} g_1 \\ g_2 \\ g_3 \end{bmatrix},$$

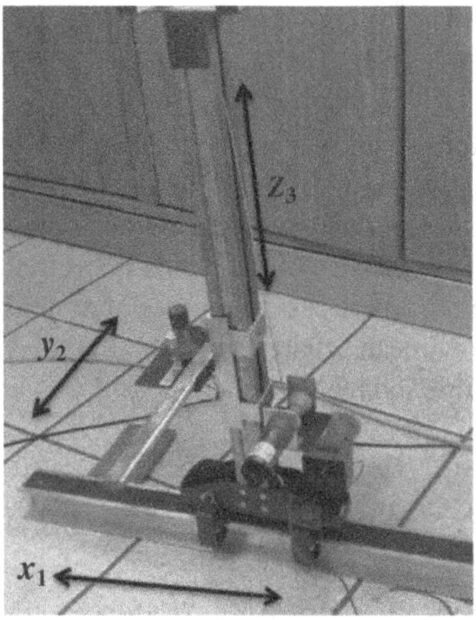

Figure 10.3 Transelevator.

where

$$m_{11} = m_1 + m_2 + m_3,$$
$$m_{22} = m_2 + m_3,$$
$$m_{33} = m_3.$$

The other parameters of $M(\chi_1)$ are zero, all the parameters of $C(\chi_1, \chi_2)$ are zero, $g_2 = (m_2 + m_3)\, g$, and the other parameters of $G(\chi_1)$ are zero. m_1, m_2, and m_3 are the masses of the links one, two, and three, respectively. $\chi_{11} = x_1$, $\chi_{12} = y_2$, and $\chi_{13} = z_3$, are the lengths of the links one, two, and three, respectively; in m, g is the acceleration gravity constant. $m_{11} = 2.9020$ kg, $m_{22} = 0.5480$ kg, $m_{33} = 0.150$ kg, $m_2 + m_3 = 0.150$ kg, and $g = 9.81$ m/s^2 are some important parameters of this robotic arm. $n_r = n_l = 0.5$, $a_r = 0.5$, and $a_l = -0.5$ are the parameters of the dead zone.

PD is given by (10.7) with parameters $K_p = \begin{bmatrix} 2000 & 0 & 0 \\ 0 & 2000 & 0 \\ 0 & 0 & 200 \end{bmatrix}$,

$K_d = \begin{bmatrix} 30 & 0 & 0 \\ 0 & 30 & 0 \\ 0 & 0 & 10 \end{bmatrix}$, $G = \begin{bmatrix} g_1 \\ g_2 \\ g_3 \end{bmatrix}$.

SM is given by (10.8) with parameters $K_p = \begin{bmatrix} 1000 & 0 & 0 \\ 0 & 1000 & 0 \\ 0 & 0 & 100 \end{bmatrix}$,

$L_s = \begin{bmatrix} 50 & 0 & 0 \\ 0 & 80 & 0 \\ 0 & 0 & 8 \end{bmatrix}$, $\lambda_s = \begin{bmatrix} 1 \\ 1 \\ 1 \end{bmatrix}$.

SSM is given by (10.12) with parameters $K_p = \begin{bmatrix} 1000 & 0 & 0 \\ 0 & 1000 & 0 \\ 0 & 0 & 100 \end{bmatrix}$,

$K_d = \begin{bmatrix} 30 & 0 & 0 \\ 0 & 30 & 0 \\ 0 & 0 & 10 \end{bmatrix}$. Since $|G(\chi_1)| \leq \overline{G} = \begin{bmatrix} 0 \\ 1.4715 \\ 0 \end{bmatrix}$ and $|d(w)| \leq \overline{d} = \begin{bmatrix} 0.25 \\ 0.25 \\ 0.25 \end{bmatrix}$, with $K = \begin{bmatrix} k_1 \\ k_2 \\ k_3 \end{bmatrix} = \begin{bmatrix} 0.25 \\ 1.73 \\ 0.25 \end{bmatrix}$, the condition given in (10.13) $\overline{G} + \overline{d} \leq K$ is satisfied; so, the regulation error of the SSM applied to the robotic arm with unknown behaviors in the dead zone and gravity is warranted to be asymptotically stable.

The comparisons for the regulator functions are shown in Figure 10.4, the position states are shown in Figure 10.5, and the velocity states are shown in Figure 10.6. Table 10.1 shows the RMSE using (10.14).

Note that the PD strategy requires bigger gains than the SSM strategy to obtain satisfactory benefits. It could cause an incorrect functioning of the PD strategy because it could yield the undesired saturation of its gains. Note that in the SM the undesired chattering is presented, it increases the RMSE, and it could damage the devices of the system. From Figures 10.4 to 10.6, it can be seen that the SSM improves both the PD and SM because the states of the first follow the trajectories of the desired constant references better than the states of the other; in addition, the inputs and regulation errors for the first are smaller than for the other. And finally, from Table 10.1, it can be shown that the SSM achieves better accuracy

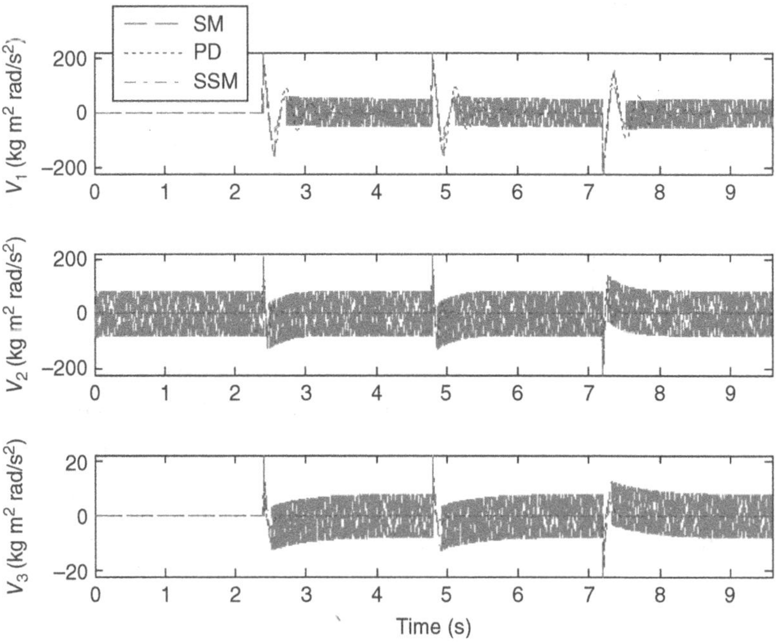

Figure 10.4 Inputs for the transelevator.

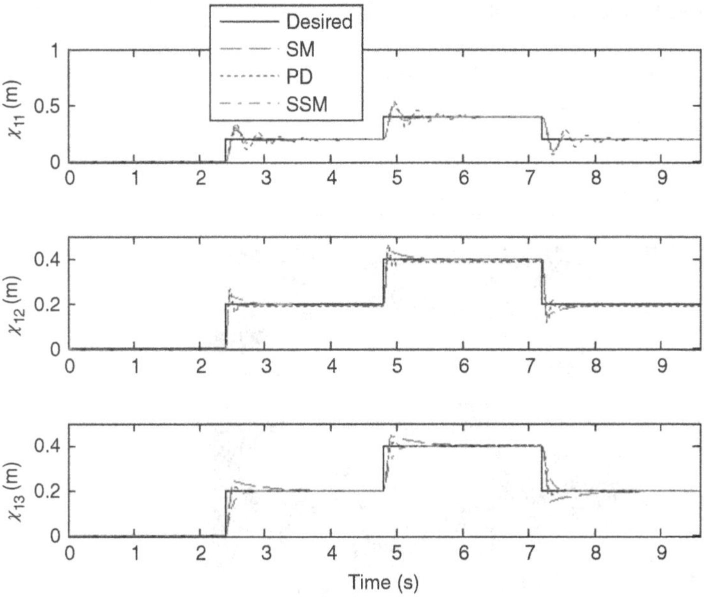

Figure 10.5 Position states for the transelevator.

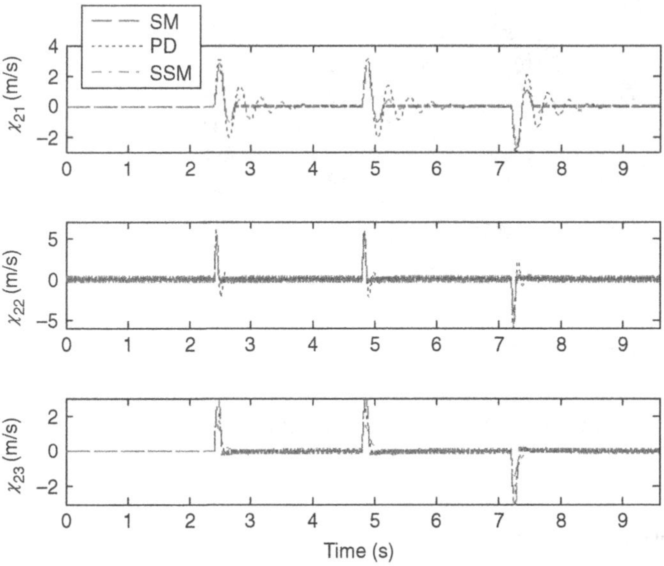

Figure 10.6 Velocity states for the transelevator.

Table 10.1 Errors for the transelevator.

Strategies	RMSE for $\tilde{\chi}$	RMSE for ν
PD	0.6077	807.9397
SM	0.4924	4380.2
SSM	0.3347	398.8813

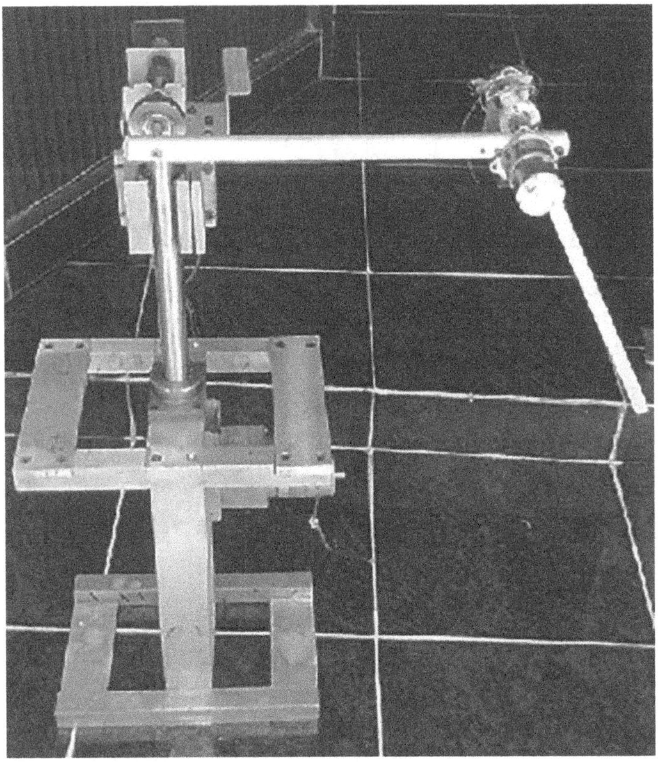

Figure 10.7 Articulated robotic arm.

when compared with both the PD and SM because the RMSE is smaller for the first than for the other.

10.4.2 Articulated Robotic Arm

Consider the articulated robotic arm shown in Figure 10.7.

The articulated robotic arm is written as (10.1), and it is detailed as follows:

$$M(\chi_1) = \begin{bmatrix} m_{11} & m_{12} & m_{13} \\ m_{21} & m_{22} & m_{23} \\ m_{31} & m_{32} & m_{33} \end{bmatrix},$$
$$C(\chi_1, \chi_2) = \begin{bmatrix} c_{11} & c_{12} & c_{13} \\ c_{21} & c_{22} & c_{23} \\ c_{31} & c_{32} & c_{33} \end{bmatrix}, \quad (10.16)$$
$$G(\chi_1) = \begin{bmatrix} g_1 \\ g_2 \\ g_3 \end{bmatrix},$$

where

$$\begin{aligned} m_{11} &= J_{13} + \left(m_2 l_{c2}^2 + m_3 l_2^2\right) C_2^2 \\ &\quad + 2m_3 l_2 l_{c3} C_2 C_{23} + m_3 l_{c3}^2 C_{23}^2, \\ m_{22} &= J_{23} + m_3 l_2^2 + m_3 l_{c3}^2 C_2^2 + m_2 l_{c2}^2 + 2m_3 l_2 l_{c3} C_3, \\ m_{33} &= J_3 + m_3 l_{c3}^2, \\ m_{23} &= J_3 + m_3 l_{c3}^2 + m_3 l_2 l_{c3} C_3, \\ m_{32} &= \left[J_3 + m_3 l_{c3}^2 + m_3 l_2 l_{c3} C_3\right]. \end{aligned}$$

The other parameters of $M(\chi_1)$ are zero.

$$\begin{aligned} c_{11} &= \left[-\left(m_2 l_{c2}^2 + m_3 l_2^2\right) S_2 C_2 - m_3 l_2 l_{c3} S_2 C_{23} \right. \\ &\quad \left. - m_3 l_2 l_{c3} C_2 S_{23} - m_3 l_{c3}^2 S_{23} C_{23}\right] \chi_{22} \\ &\quad + (-m_3 l_2 l_{c3} C_2 S_{23} - m_3 l_{c3}^2 S_{23} C_{23}) \chi_{23}, \\ c_{12} &= \left[-\left(m_2 l_{c2}^2 + m_3 l_2^2\right) S_2 C_2 - m_3 l_2 l_{c3} S_2 C_{23} \right. \\ &\quad \left. - 2m_3 l_2 l_{c3} C_2 S_{23} - m_3 l_{c3}^2 S_{23} C_{23}\right] \chi_{21}, \\ c_{13} &= (-m_3 l_2 l_{c3} C_2 S_{23} - m_3 l_{c3}^2 S_{23} C_{23}) \chi_{21}, \\ c_{22} &= -m_3 l_2 l_{c3} S_3 \chi_{23}, \\ c_{23} &= -m_3 l_2 l_{c3} S_3 \chi_{23} - m_3 l_2 l_{c3} S_3 \chi_{22}, \\ c_{32} &= -m_3 l_2 l_{c3} S_3 \chi_{23}. \end{aligned}$$

The other parameters of $C(\chi_1, \chi_2)$ are zero.

$$g_2 = (m_2 l_{c2} + m_3 l_2)\, g C_2 + m_3 g l_{c3} C_{23},$$
$$g_3 = m_3 g l_{c3} C_{23}.$$

The other parameters of $G(\chi_1)$ are zero. m_1, m_2, and m_3 are the masses of the links one, two, and three, respectively; in kg, $\chi_{11} = \theta_1$, $\chi_{12} = \theta_2$, and $\chi_{13} = \theta_3$, are the angles of the joints one, two, and three, respectively; in rad, g is the acceleration gravity constant, J_1, J_2, and J_3 are the inertias in kg m^2. $C_2 = \cos(\chi_{21})$, $C_3 = \cos(\chi_{12})$, $C_{23} = \cos(\chi_{21} + \chi_{12})$, $S_2 = \sin(\chi_{21})$, $S_3 = \sin(\chi_{12})$, and $S_{23} = \sin(\chi_{21} + \chi_{12})$. $m_2 = 0.34$ kg, $m_3 = 0.24$ kg, $l_2 = 0.293$ m, $l_3 = 0.28$ m, $l_{c2} = \frac{l_2}{2}$, $l_{c3} = \frac{l_3}{2}$, $J_{13} = J_1 + J_2 + J_3 = 0.04624$ kg m^2, $J_{23} = J_2 + J_3 = 0.02545$ kg m^2, $J_3 = 0.01273$ kg m^2, and $g = 9.81$ m/s^2 are some important parameters of this robotic arm. $n_r = n_l = 0.5$, $a_r = 0.5$, and $a_l = -0.5$ are the parameters of the dead zone.

PD is given by (10.7) with parameters $K_p = \begin{bmatrix} 500 & 0 & 0 \\ 0 & 1000 & 0 \\ 0 & 0 & 1000 \end{bmatrix}$,

$$K_d = \begin{bmatrix} 10 & 0 & 0 \\ 0 & 10 & 0 \\ 0 & 0 & 10 \end{bmatrix}, G = \begin{bmatrix} g_1 \\ g_2 \\ g_3 \end{bmatrix}.$$

SM is given by (10.8) with parameters $K_p = \begin{bmatrix} 250 & 0 & 0 \\ 0 & 500 & 0 \\ 0 & 0 & 500 \end{bmatrix}$,

$$L_s = \begin{bmatrix} 20 & 0 & 0 \\ 0 & 27 & 0 \\ 0 & 0 & 38 \end{bmatrix}, \lambda_s = \begin{bmatrix} 1 \\ 1 \\ 1 \end{bmatrix}.$$

SSM is given by (10.12) with parameters $K_p = \begin{bmatrix} 250 & 0 & 0 \\ 0 & 500 & 0 \\ 0 & 0 & 500 \end{bmatrix}$,

$K_d = \begin{bmatrix} 10 & 0 & 0 \\ 0 & 10 & 0 \\ 0 & 0 & 10 \end{bmatrix}$. Since $|G(\chi_1)| \le \overline{G} = \begin{bmatrix} 0 \\ 1.1783 \\ 0.329616 \end{bmatrix}$ and $|d(w)| \le \overline{d} = \begin{bmatrix} 0.25 \\ 0.25 \\ 0.25 \end{bmatrix}$, with $K = \begin{bmatrix} k_1 \\ k_2 \\ k_3 \end{bmatrix} = \begin{bmatrix} 0.25 \\ 1.43 \\ 0.58 \end{bmatrix}$, the condition given in (10.13) $\overline{G} + \overline{d} \leq K$ is satisfied; so, the regulation error of the SSM applied to the robotic arm with unknown behaviors in the dead zone and gravity is warranted to be asymptotically stable.

The comparisons for the regulator functions are shown in Figure 10.8, the position states are shown in Figure 10.9, and the velocity states are shown in Figure 10.10. Table 10.2 shows the RMSE using (10.14).

Note that the PD strategy requires bigger gains than the SSM strategy to obtain satisfactory benefits. It could cause an incorrect functioning of the PD strategy because it could yield the undesired saturation of its gains. Note that in the SM the undesired chattering is presented, it increases the RMSE, and it could damage the devices of the system. From Figures 10.8 to 10.10, it can be seen that the SSM improves both the PD and SM because the states of the first follow the trajectories of the desired constant references better than the states of the other; in addition, in the first the inputs and regulation errors are smaller than in the other. And finally, from Table 10.2, it can be shown that the SSM achieves better accuracy when compared with both the PD and SM because the RMSE is smaller for the first than for the other.

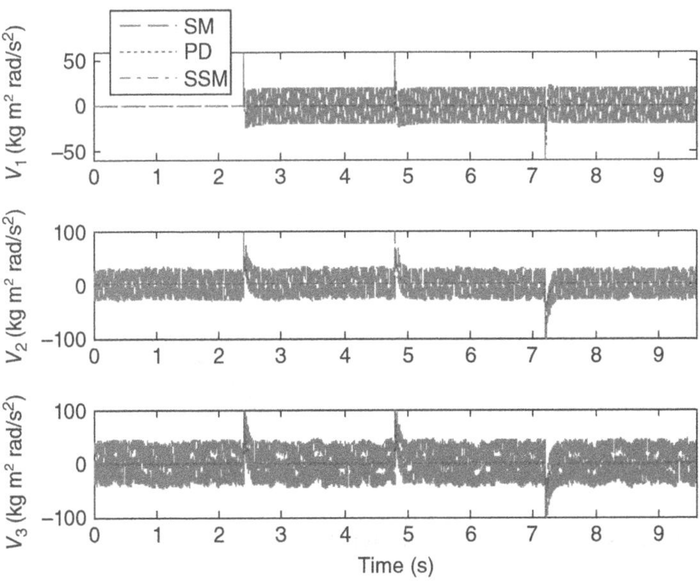

Figure 10.8 Inputs for the articulated robotic arm.

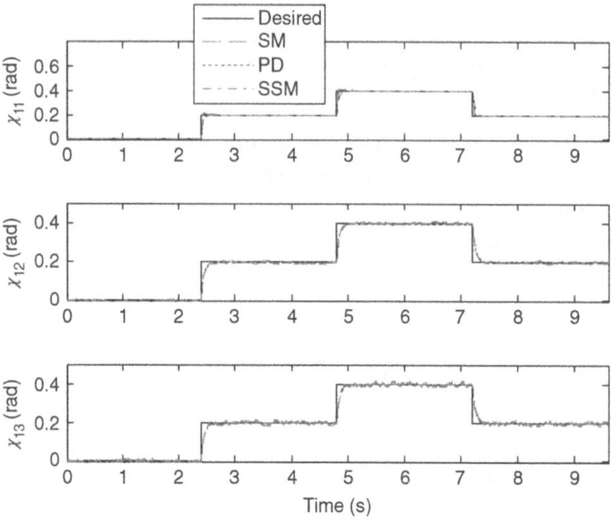

Figure 10.9 Position states for the articulated robotic arm.

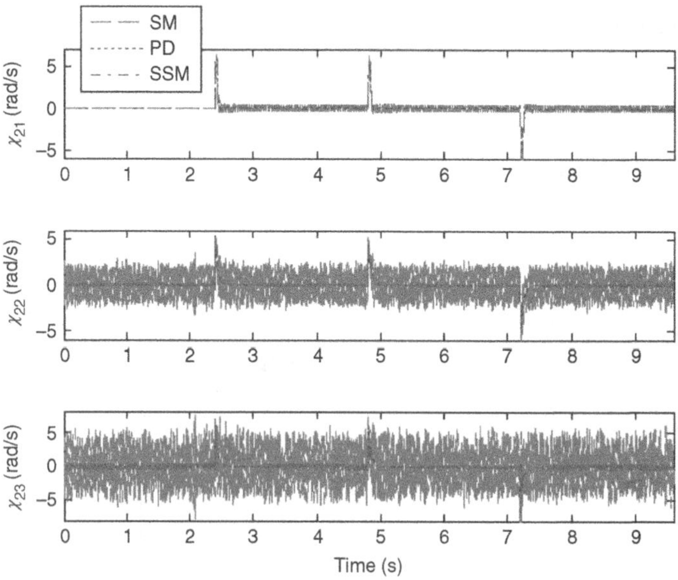

Figure 10.10 Velocity states for the articulated robotic arm.

Table 10.2 Errors for the articulated robotic arm.

Strategies	RMSE for $\tilde{\chi}$	RMSE for ν
PD	0.3592	43.4465
SM	4.4764	1281.6
SSM	0.3282	42.9560

Remark 10.4: The comparisons show that the SSM improved both the PD and SM for the regulation of two robotic arms with unknown behaviors in the dead zone and gravity; this is because the PD and SM are recommended for robotic arms that do not consider the dead-zone terms.

10.5 Concluding Remarks

In this chapter, an SM regulator is presented for the regulation of robotic arms with unknown behaviors of the dead zone and gravity. The comparisons showed that the recommended strategy achieves better performance compared with both the PD regulator with gravity compensation and the SM regulator for the regulation of two robotic arms. The comparisons illustrate the viability, efficiency, and potential of the approach especially important in robotics. The recommended strategy could also be applied to cranes or pendulum processes. As a future research, the recommended strategy will be improved by considering that some parameters of the regulator are approximated by the intelligent processes.

References

1 Abrir S., Xie W. F., Su C. Y. (2007), Adaptive tracking of nonlinear systems with non-symmetric dead-zone input, Automatica, 43, 522–530.
2 Perez-Cruz J. H., Rubio J. J., Linares-Flores J., Rangel E. (2015), Control of uncertain plants with unknown deadzone via differential neural networks, IEEE Latin America Transactions, 13 (7), 2085–2093.
3 Deng W., Yao J., Ma D. (2015), Robust adaptive asymptotic tracking control of a class of nonlinear processes with unknown input dead-zone, Journal of the Franklin Institute, 352 (12), 5686–5707.
4 Han S. I., Lee J. (2016), Finite-time sliding surface constrained control for a robot manipulator with an unknown dead zone and disturbance, ISA Transactions, 65, 307–318.

5 Hou M., Zhang Z., Deng Z., Duan G. (2016), Global robust finite-time stabilisation of unknown pure-feedback systems with input dead-zone non-linearity, IET Control Theory and Applications, 10 (2), 234–243.
6 Wang F., Liu Z., Zhang Y., Chen B. (2016), Distributed adaptive coordination control for uncertain nonlinear multi-agent systems with dead-zone input, Journal of the Franklin Institute, 353, 2270–2289.
7 Zuo Z., Li X., Shi Z., (2015), L1 adaptive control of uncertain gear transmission servo systems with deadzone nonlinearity, ISA Transactions, 58, 67–75.
8 Cheong J. Y., Han S. I., Lee J. M. (2013), Adaptive fuzzy dynamic surface sliding mode position control for a robot manipulator with friction and deadzone, Mathematical Problems in Engineering, 2013, 1–15.
9 Park S. H., Han S. I. (2011), Robust-tracking control for robot manipulator with deadzone and friction using backstepping and RFNN controller, IET Control Theory and Applications, 5 (12), 1397–1417.
10 He W., David A. O., Yin Z., Sun C. (2016), Neural network control of a robotic manipulator with input deadzone and output constraint, IEEE Transactions on Systems, Man and Cybernetics: Systems, 46 (6), 759–770.
11 Zhou Q., Shi P., Tian Y., Wang M. (2015), Approximation-based adaptive tracking control for mimo nonlinear systems with input saturation, IEEE Transactions on Cybernetics, 45 (10), 2119–2128.
12 He W., Chen Y., Yin Z., (2016), Adaptive neural network control of an uncertain robot with full-state constraints, IEEE Transactions on Cybernetics, 46 (3), 620–629.
13 He W., Dong Y., Sun C. (2016), Adaptive neural impedance control of a robotic manipulator with input saturation, IEEE Transactions on Systems, Man and Cybernetics: Systems, 46 (3), 334–344.
14 Liu M., Zhang L., Shi P., Reza Karimi H. (2014), Robust control of stochastic systems against bounded disturbances with application to flight control, IEEE Transactions on Industrial Electronics, 61 (3), 1504–1515.
15 Liu M., Zhang L., Shi P., Reza Karimi H. (2012), State feedback control against sensor faults for Lipschitz nonlinear systems via new sliding mode observer techniques, IEEE Conference on Decision and Control and European Control Conference, 7635–7640.
16 Liu M., Shi P., Zhang L., Zhao X. (2011), Fault-tolerant control for nonlinear Markovian jump systems via proportional and derivative sliding mode observer technique, IEEE Transactions on Circuits and Systems-I: Regular Papers, 58 (11), 2755–2764.
17 Kao Y., Xie J., Wang C., Reza Karimi H. (2015), A sliding mode approach to H∞ non-fragile observer-based control design for uncertain Markovian neutral-type stochastic systems, Automatica, 52, 218–226.

18 Kao Y., Xie J., Wang C., Reza Karimi H. (2016), Observer-based H∞ sliding mode controller design for uncertain stochastic singular time-delay systems, Circuits Systems and Signal Processing, 35, 63–77.

19 Liu Y., Niu Y., Zou Y., Reza Karimi H., (2015), Adaptive sliding mode reliable control for switched systems with actuator degradation, IET Control Theory and Applications, 9 (8), 1197–1204.

20 Wang B., Shi P., Reza Karimi H. (2014), Fuzzy sliding mode control design for a class of disturbed systems, Journal of the Franklin Institute, 351, 3593–3609.

21 Li H., Shi P., Yao D., Wu L. (2016), Observer-based adaptive sliding mode control for nonlinear Markovian jump systems, Automatica, 64, 133–142.

22 Li B., Du H., Li W. (2016), Fault-tolerant control of electric vehicles within-wheel motors using actuator-grouping sliding mode controllers, Mechanical Systems and Signal Processing, 72-73, 462–485.

23 Xiao X., Zhang Y., Wang J., Du H. (2016), New adaptive sliding mode observer design for sensorless control of PMSM in electric vehicle drive system, International Journal of Smart Sensing and Intelligent Systems, 9 (1), 377–396.

24 Zhou Q., Shi P., Xu S., Li H. (2013), Observer-based adaptive neural network control for nonlinear stochastic systems with time delay, IEEE Transactions on Neural Networks and Learning Systems, 24 (1), 71–80.

25 Izaguirre C., Munoz A. J., Sanchez A., Parra V., Castillo P. (2016), Attitude control of quadrotors based on fractional sliding modes: theory and experiments, IET Control Theory and Applications, 10 (7), 825–832.

26 Nojavanzadeh D., Badamchizadeh M. (2016), Adaptive fractional-order non-singular fast terminal sliding mode control for robot manipulators, IET Control Theory and Applications, 10 (13), 1565–1572.

27 Zhang S., Lei M., Dong Y., He W. (2016), Adaptive neural network control of coordinated robotic manipulators with output constraint, IET Control Theory and Applications, 10 (17), 2271–2278.

28 Lewis F. L., Dawson D. M., Abdallah C. T. (2004), Control of Robot Manipulators, Theory and Practice, Marcel Dekker, Inc.

29 Spong M. W., Hutchinson S., Vidyasagar M., (2006), Robot Modeling and Control, John Wiley & Sons.

30 Rubio J. J. (2015), Adaptive least square control in discrete time of robotic arms, Soft Computing, 19 (12), 3665–3676.

31 Rubio J. J. (2017), Sliding mode control of robotic arms with deadzone, IET Control Theory and Applications, 11 (8), 1214–1221.

Index

c

chattering 50, 54, 55, 118, 121, 128, 135, 136, 145, 192, 196, 200, 205
convergence 26, 27, 29, 30, 45, 53, 56, 94, 98, 114, 118, 120, 135, 148, 149, 168, 178, 197

d

dead-zone 25, 46, 139–141, 143–157, 173–179, 188, 191–197, 199, 200, 204, 205, 207

e

eigenvalue 29, 35, 42, 53, 56, 98, 104, 111, 118, 122, 127, 134, 165, 167
electric vehicle 94, 108–114, 192
electricity generator 108, 109, 139–141, 143–157
error 1, 26, 30, 51–53, 55, 56, 76–78, 94, 96–98, 104, 105, 111, 112, 114, 118–120, 135, 140, 142, 148, 149, 156, 161, 177–183, 185–187, 194–198, 200, 202, 205, 207

g

gas turbine 118, 123, 124, 128, 129, 135
generator 94, 101, 102, 108–110, 139, 140, 142, 145, 146, 148, 149, 157
geometric 1–3, 11, 14, 15, 18, 20–22, 75, 78, 82, 87, 89, 93

h

hybrid 49–58, 67, 68, 72
hydraulic 45, 72, 84, 85, 87, 93, 114, 136, 173, 191

l

linear 2, 25, 26, 93, 110, 119, 165, 166, 173
Lyapunov 29, 45, 50–52, 54, 59, 67, 76, 77, 82, 87, 95, 97, 114, 118, 120, 122, 127, 134, 136, 140, 161, 192, 196

m

mathematical model 8, 25, 35, 38, 58, 59, 67, 81, 87, 95, 103, 109, 140, 142, 144, 146, 147, 174, 176, 192
mechanical 1, 2, 8, 13–15, 22, 136, 173, 191

motor 25, 38, 39, 54, 55, 79, 84, 101, 102, 108–110, 149, 157, 163, 173, 191

n

nonlinear 25, 26, 96, 102, 107, 117–122, 128, 129, 134–136, 161, 173, 174, 178, 191, 193
nonlinearities 1, 139, 140, 142, 146–149, 157, 173, 191

o

observer 1, 25, 49–56, 58, 59, 65, 67, 68, 70, 72, 75, 76, 93–98, 103–107, 111–114, 192, 197

p

process 1–3, 5, 6, 8–10, 13–17, 19–22, 25–29, 31, 32, 35, 38, 40, 42, 45, 49, 50–56, 59–72, 75–82, 84–87, 89, 93–99, 102, 103, 105–107, 109, 112–114, 117–125, 128–132, 134–136, 161–167, 169, 173, 174, 176–180, 185, 188, 191, 192, 195–197, 207
pendulum 149, 157, 161–164, 166, 168–170, 173, 174, 176, 177, 179, 180, 182–185, 188, 207
perturbation 1–3, 5– 8, 10, 13–17, 19, 20, 22, 25–28, 30–32, 35, 36, 38, 41–43, 49–56, 58, 59, 61, 62, 65– 67, 69, 70, 72, 75–79, 81, 82, 84, 87–89, 93–99, 102–104, 106, 107, 109–111, 113, 117, 119, 121–125, 130, 132, 135, 139, 142, 145, 146, 148, 149, 157, 161, 162, 166–169, 196
perturbation attenuation 1–3, 5–8, 10, 11, 17, 18, 22, 25–30, 35, 38, 42, 45, 49–51, 54, 56, 72, 75, 76, 78, 79, 82, 84, 87, 89, 93, 94, 117, 136, 161, 162, 165, 166, 169
perturbation estimation 49–51, 53–55, 72, 75, 76, 93, 94, 96, 98, 107, 111, 114, 117–119, 121, 122, 128, 129, 134–136
plotter 26, 38, 39, 42–45, 51, 56, 58, 59, 61–64, 72
polynomial 7, 165, 167, 173
position 1, 13, 25, 101, 103, 111, 139, 140, 147, 149, 153, 156, 157, 162, 163, 177, 181, 182, 184, 185, 192, 194–196, 198, 200, 201, 205, 206
proportional derivative 139, 140, 142, 147, 148, 157, 165, 173, 174, 177, 192

r

regulation 30, 75, 117, 139, 140, 147–149, 157, 162, 165, 169, 173, 174, 177, 179, 188, 191, 192, 194–198, 200, 205, 207
regulator 1–8, 10, 11, 13–22, 25–31, 33, 35, 38, 40–42, 45, 49–51, 53–56, 58, 59, 61, 62, 66, 67, 69, 72, 75–78, 81, 82, 84, 87–89, 93, 94, 96, 99, 107, 109, 117, 139, 140, 142, 147–149, 157, 161, 162, 165–169, 161, 162, 165, 166, 166, 167, 173–179, 181, 182, 185, 188, 191–197, 200, 205, 207
robotic arm 162, 163, 172, 174, 176, 184, 191, 192, 195–197, 199, 200, 202–207
root mean square error 30, 56, 78, 99, 123, 149, 167, 179, 197

s

sliding mode 49, 50, 54, 75–77, 79, 82, 84, 87, 89, 118, 139, 147, 157, 173, 191, 195, 196
stability 22, 26, 27, 29, 30, 45, 50–56, 76–78, 94, 95, 97, 98, 114, 161, 162, 165, 167, 173, 174, 176, 192, 194, 196

state 1–3, 8, 9, 13, 14, 16, 20, 21, 25–27, 29, 31, 36, 38, 42, 43, 49, 51–53, 55, 56, 58–61, 63–70, 75–78, 80–83, 85–88, 93–99, 102–105, 107, 109, 111, 112, 114, 142, 147–153, 155, 156, 161, 163, 164, 166, 169, 170, 175, 181, 182, 185, 186, 193, 196, 197, 200, 201, 205, 206
structure at infinity 1, 2, 6, 7, 10, 17, 22, 25–28, 30, 33, 35–38, 40–45, 93
structure regulator 25–30, 33, 35, 38, 40–42, 45
subspace 2–5, 11, 18
suspension 51, 64–72
system 34, 41, 157, 173, 175, 178, 179, 191–193, 196, 197, 200, 205

t

thermal 1–3, 15, 16, 20–22, 45, 72, 79–81, 114, 136
three phase circuit 26, 30–32, 34–38, 45
transfer function 3, 6, 10, 17, 26–28, 33, 40, 166, 168

v

variable 7, 31, 64, 65, 93, 95, 99, 102, 117–123, 125, 128, 129, 132, 134, 135, 182, 185, 192
vehicle 64, 65, 93, 94, 109
velocity 1, 39, 56, 110, 162, 177, 179, 181–183, 185, 187, 192, 194–198, 200, 201, 205, 206

w

wind turbine 94, 99, 100–107, 114, 139, 173